17th Edition
IEE Wiring Regulations:
Inspection, Testing and Certification

By the same author

17th Edition IEE Wiring Regulations: Design and Verification of Electrical Installations, ISBN 978-0-0809-6914-5
17th Edition IEE Wiring Regulations: Explained and Illustrated, ISBN 978-0-7506-8720-1
Electric Wiring: Domestic, ISBN 978-0-7506-8735-5
PAT: Portable Appliance Testing, ISBN 978-0-0809-6919-0
Wiring Systems and Fault Finding, ISBN 978-0-7506-8734-8
Electrical Installation Work, ISBN 978-0-7506-8733-1

17th Edition
IEE Wiring Regulations:
Inspection, Testing and Certification

Seventh Edition

Brian Scaddan, IEng, MIET

AMSTERDAM • BOSTON • HEIDELBERG • LONDON • NEW YORK
OXFORD • PARIS • SAN DIEGO • SAN FRANCISCO
SINGAPORE • SYDNEY • TOKYO
Newnes is an imprint of Elsevier

Newnes is an imprint of Elsevier
The Boulevard, Langford Lane, Kidlington, Oxford OX5 1GB, UK
225 Wyman Street, Waltham, MA 02451, USA

First edition 1996
Second edition 1998
Third edition 2001
Fourth edition 2002
Fifth edition 2005
Sixth edition 2008
Seventh edition 2011

Notice
No responsibility is assumed by the publisher for any injury and/or damage to persons or property as a matter of products liability, negligence or otherwise, or from any use or operation of any methods, products, instructions or ideas contained in the material herein. Because of rapid advances in the medical sciences, in particular, independent verification of diagnoses and drug dosages should be made

British Library Cataloguing in Publication Data
A catalogue record for this book is available from the British Library

Library of Congress Cataloging-in-Publication Data
A catalog record for this book is availabe from the Library of Congress

ISBN: 978-0-08-096610-6

For information on all Newnes publications
visit our web site at books.elsevier.com

Printed and bound in Italy
11 12 13 14 15 10 9 8 7 6 5 4 3 2 1

Working together to grow
libraries in developing countries

www.elsevier.com | www.bookaid.org | www.sabre.org

ELSEVIER   BOOK AID International   Sabre Foundation

Contents

Preface

As a bridge between the 17th Edition course (C&G 2382-10) and the Design, Erection and Verification course (C&G 2391-20), the author, in association with the City & Guilds and Donald Malcolm Consultants, was involved in the development of the Inspection, Testing and Certification courses (C&G 2392-10 and 2391-10). The 2392-10 covers the requirements for Initial Verification and the 2391-10 both Initial Verification and Periodic Inspection and Testing.

This book has been revised to serve as an accompaniment to these new schemes and has been brought fully up-to-date with the 17th Edition Wiring Regulations. It is also a useful addition to the reference library of contracting electricians and candidates studying for the C&G 2382 and 2391-20 qualifications.

Brian Scaddan, February 2011

Material on Part P in Chapter 1 is taken from *Building Regulations Approved Document P: Electrical Safety – Dwelling*, P1 Design and installation of electrical installations (The Stationery Office, 2006) ISBN 9780117036536. © Crown copyright material is reproduced with the permission of the Controller of HMSO and Queen's Printer for Scotland.

Introduction

THE IEE WIRING REGULATIONS BS 7671

Before we embark on the subject of inspection and testing, it is, perhaps, wise to refresh our memories with regards to one or two important topics from the 17th edition (BS 7671 2008).

Clearly, the protection of persons and livestock from shock and burns, etc. and the prevention of damage to property are priorities. In consequence, therefore, thorough inspection and testing of an installation and subsequent remedial work where necessary will significantly reduce the risks.

So let us start with electric shock, that is the passage of current through the body of such magnitude as to have significant harmful effects. Figure 0.1 illustrates the generally accepted effects of current passing through the human body.

How then are we at risk of electric shock, and how do we protect against it?

There are two ways in which we can be at risk:

1. **T**ouching live parts of equipment or systems that are intended to be live.
2. **T**ouching conductive parts which are not meant to be live, but have become live due to a fault.

The conductive parts associated with the second of these can either be metalwork of electrical equipment and accessories (Class I) and that of electrical wiring systems such as metallic conduit and trunking, etc. called *exposed conductive parts*, or other metalwork such as pipes, radiators, girders, etc. called *extraneous conductive parts*.

Let us now consider how we may protect against electric shock from whatever source.

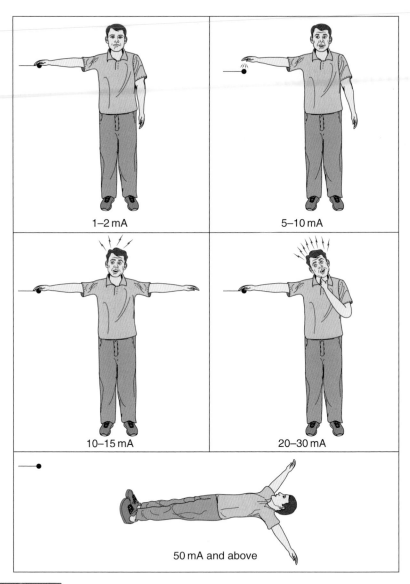

FIGURE 0.1 Shock levels.

1–2 mA	Barely perceptible, no harmful effects
5–10 mA	Throw off, painful sensation
10–15 mA	Muscular contraction, can't let go
20–30 mA	Impaired breathing
50 mA and above	Ventricular fibrillation and death

PROTECTION AGAINST SHOCK FROM BOTH TYPES OF CONTACT

One method of achieving this is by ensuring that the system voltage does not exceed extra low (50 V ac, 120 V ripple-free dc), and that all associated wiring, etc. is separated from all other circuits of a higher voltage and Earth. Such a system is known as a *separated extra low voltage (SELV)*. If a SELV system exceeds 25 V ac, 60 V ripple-free dc, then extra protection must be provided by barriers, enclosures and insulation.

BASIC PROTECTION

Apart from SELV, how can we prevent danger to persons and livestock from contact with intentionally live parts? Clearly we must minimize the risk of such contact, and this may be achieved in one or more of the following ways:

1. Insulate any live parts.
2. Ensure that any uninsulated live parts are housed in suitable enclosures and/or are behind barriers.
3. Place obstacles in the way. (This method would only be used in areas where skilled and/or authorized persons were involved.)
4. Placing live parts out of reach. (Once again, only used in special circumstances, e.g. live rails of overhead travelling cranes.)

A residual current device (RCD) may be used as additional protection to any of the other measures taken, provided that it is rated at 30 mA or less and has an operating time of not more than 40 ms at a test current of five times its operating current.

It should be noted that RCDs are not the panacea for all electrical ills, they can malfunction, but they are a valid and effective back-up to the other methods. They must not be used as the sole means of protection.

FAULT PROTECTION

How can we protect against shock from contact with unintentionally live, exposed or extraneous conductive parts whilst touching earth, or from contact between unintentionally live exposed and/or extraneous conductive parts? The most common method is by *protective earthing, protective equipotential bonding and automatic disconnection in case of a fault.*

All extraneous conductive parts are connected with a main protective bonding conductor and connected to the main earthing terminal, and all exposed conductive parts are connected to the main earthing terminal by the circuit protective conductors (cpc). Add to this overcurrent protection that will operate fast enough when a fault occurs and the risk of severe electric shock is significantly reduced.

Other means of fault protection may be used, but are less common and some require very strict supervision.

USE OF CLASS II EQUIPMENT

Often referred to as double-insulated equipment, this is typical of modern appliances where there is no provision for the connection of a cpc. This does not mean that there should be no exposed conductive parts and that the casing of equipment should be of an insulating material; it simply indicates that live parts are so well insulated that faults from live to conductive parts cannot occur.

NON-CONDUCTING LOCATION

This is basically an area in which the floor, walls and ceiling are all insulated. Within such an area there must be no protective conductors, and socket outlets will have no earthing connections.

It must not be possible simultaneously to touch two exposed conductive parts, or an exposed conductive part and an extraneous conductive part. This requirement clearly prevents shock current passing through a person in the event of an earth fault, and the insulated construction prevents shock current passing to earth.

EARTH-FREE LOCAL EQUIPOTENTIAL BONDING

This is in essence a Faraday cage, where all metals are bonded together but *not* to earth. Obviously, great care must be taken when entering such a zone in order to avoid differences in potential between inside and outside.

The areas mentioned in this and the previous method are very uncommon. Where they do exist, they should be under constant supervision to ensure that no additions or alterations can lessen the protection intended.

ELECTRICAL SEPARATION

This method relies on a supply from a safety source such as an isolating transformer to BS EN 61558-2-6 which has no earth connection on the secondary side. In the event of a circuit that is supplied from a source developing a live fault to an exposed conductive part, there would be no path for shock current to flow (see Figure 0.2).

Once again, great care must be taken to maintain the integrity of this type of system, as an inadvertent connection to earth, or interconnection with other circuits, would render the protection useless.

Additional protection by RCDs is a useful back-up to other methods of shock protection.

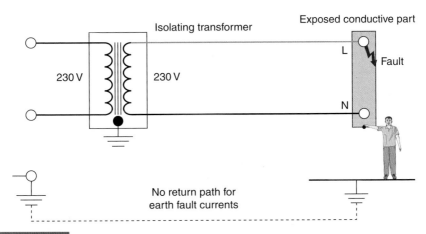

FIGURE 0.2 Electrical separation

Table 0.1 IP Codes

First numeral:	Mechanical protection
0	No protection of persons against contact with live or moving parts inside the enclosure. No protection of equipment against ingress of solid foreign bodies.
1	Protection against accidental or inadvertent contact with live or moving parts inside the enclosure by a large surface of the human body, for example a hand, but not protection against deliberate access to such parts.
2	Protection against ingress of large solid foreign bodies. Protection against contact with live or moving parts inside the enclosure by fingers. Protection against ingress of medium-size solid foreign bodies.
3	Protection against contact with live or moving parts inside the enclosures by tools, wires or such objects of thickness greater than 2.5mm. Protection against ingress of small foreign bodies.
4	Protection against contact with live or moving parts inside the enclosure by tools, wires or such objects of thickness greater than 1mm. Protection against ingress of small-size solid foreign bodies.
5	Complete protection against contact with live or moving parts inside the enclosure. Protection against harmful deposits of dust. The ingress of dust is not totally prevented, but dust cannot enter in an amount sufficient to interfere with satisfactory operation of the equipment enclosed.
6	Complete protection against contact with live or moving parts inside the enclosures. Protection against ingress of dust.
Second numeral:	**Liquid protection**
0	No protection.
1	Protection against drops of condensed water. Drops of condensed water falling on the enclosure shall have no harmful effect.
2	Protection against drops of liquid. Drops of falling liquid shall have no harmful effect when the enclosure is tilted at any angle up to 15° from the vertical.
3	Protection against rain. Water falling in rain at an angle equal to or smaller than 60° with respect to the vertical shall have no harmful effect.
4	Protection against splashing. Liquid splashed from any direction shall have no harmful effect.
5	Protection against water jets. Water projected by a nozzle from any direction under stated conditions shall have no harmful effect.

First numeral:	Mechanical protection
6	Protection against conditions on ships' decks (deck with watertight equipment). Water from heavy seas shall not enter the enclosures under prescribed conditions.
7	Protection against immersion in water. It must not be possible for water to enter the enclosure under stated conditions of pressure and time.
8	Protection against indefinite immersion in water under specified pressure. It must not be possible for water to enter the enclosure.
X	Indicates no specified protection.

The use of enclosures is not limited to protection against shock from contact with live parts; they clearly provide protection against the ingress of foreign bodies and moisture. In order to establish to what degree an enclosure can resist such ingress, reference to the Index of Protection (IP) code (BS EN 60529) should be made. Table 0.1 illustrates part of the IP code.

The most commonly quoted IP codes in the 17th edition are IPXXB or IP2X, and IPXXD or IP4X. The X denotes that protection is not specified, not that there is no protection. For example, an enclosure that was to be immersed in water would be classified IPX8, there would be no point using the code IP68.

Note

IPXXB denotes protection against finger contact only.
IPXXD denotes protection against penetration by 1 mm diameter wires only.

CHAPTER 1

An Overview

Important terms/topics covered by this chapter.

- Statutory and Non-Statutory Regulations
- Electrical systems
- The Building Regulations Part 'P'
- Instruments

By the end of this chapter the reader should,

- be aware of the Statutory and Non-Statutory Regulations that are relevant to installation work,
- know the range of instruments required,
- know the requirements regarding the use and performance of test equipment.

So, here you are outside the premises, armed with lots of test instruments, a clipboard, a pad of documents that require completing, the IEE Regulations, Guidance Notes 3 and an instruction to carry out an inspection and test of the electrical installation therein. Dead easy, you've been told, piece of cake, just poke about a bit, 'Megger' the wiring, write the results down, sign the test certificate and you should be onto the next job within the hour!

Oh! If only it were that simple! What if lethal defects were missed by just 'poking about'? What if other tests should have been carried out which may have revealed serious problems? What if things go wrong after you have signed to say all is in accordance with the Regulations? What if you were not actually competent to carry out the inspection and test in the first place? What if … and so on, the list is endless. Inspection, testing and certification is a serious and, in many instances, a complex matter, so let us wind the clock back to the point at which you were about to enter the premises to carry out your tests, and consider the implications of carrying out an inspection and test of an installation.

IEE Wiring Regulations. DOI: 10.1016/B978-0-08-096610-6.10001-6

What are the legal requirements in all of this? Where do you stand if things go wrong? What do you need to do to ensure compliance with the law?

It is probably best at this point to consider the types of Inspection and Test that need to be conducted and the certification required.

There are two types:

1. Initial Verification.
2. Periodic Inspection and Testing.

Initial Verification is required for new work and alterations and additions (covered in City and Guilds 2392-10).

Periodic Inspection and Testing is required for existing installations (this and Initial Verification are covered in City and Guilds 2391-10).

The certification required for (1) (above) is an Electrical Installation Certification (EIC).

The certification required for (2) (above) is a Periodic Inspection Report (PIR). This could be referred to as a Condition Report.

Both must be accompanied by a schedule of test results and a schedule of inspections.

In the case of an addition or simple alteration that does not involve the installation of a new circuit (e.g. a spur from a ring final circuit), tests must be conducted but the certification required is a Minor Electrical Installation Works Certificate (MEIWC).

These are all covered in greater detail in Chapter 16.

STATUTORY AND NON-STATUTORY REGULATIONS

The statutory regulations that apply to electrical work are:

- The Health and Safety at Work Etc. Act (HSWA).
- The Electricity at Work Regulations (EAWR).
- The Building Regulations Part 'P' (applicable to domestic installations).

Non-statutory regulations include such documents as BS 7671:2008 and associated guidance notes, Guidance Note GS 38 on test equipment, etc.

The IEE Wiring Regulations (BS 7671:2008) and associated guidance notes are *not* statutory documents, they can, however, be used in a court of law to prove compliance with statutory requirements such as the Electricity at Work Regulations (EAWR) 1989, which cover all work activity associated with electrical systems. A list of other statutory regulations is given in Appendix 2 of the IEE Regulations. However, it is the EAWR that are most closely associated with BS 7671, and as such it is worth giving some areas a closer look.

In the EAWR there are 33 Regulations in all, 12 of which deal with the special requirements of mines and quarries, one which deals with extension outside Great Britain, and three which deal with effectively exemptions. We are only concerned with the first 16 Regulations, and Regulation 29, the defence regulation, which we shall come back to later. Let us start then with a comment on the meaning of *electrical systems and equipment*.

ELECTRICAL SYSTEMS AND EQUIPMENT

According to the EAWR, electrical systems and equipment can encompass anything from power stations to torch or wrist-watch batteries. A battery may not create a shock risk, but may cause burns or injury as a result of attempting to destroy it by fire, whereby explosions may occur. A system can actually include the source of energy, so a test instrument with its own supply, for example a continuity tester, is a system in itself, and a loop impedance tester, which requires an external supply source, becomes part of the system into which it is connected. From the preceding comments it will be obvious then that, in broad terms, if something is electrical, it is or is part of an electrical system. So, where does responsibility lie for any involvement with such a system?

The EAWR requires that every employer, employee and self-employed person be responsible for compliance with the Regulations with regards to matters within their control, and as such are known as *duty holders*. Where then do you stand as the person about to conduct an inspection and test of an installation? Most certainly, you are a duty holder in that you have control of the installation in so far as you will ultimately pass the installation as safe or make recommendations to ensure its safety. You also have control of the test instruments which, as already stated, are systems in themselves, and control of the installation whilst testing is being carried out.

Any breach of the Regulations may result in prosecution, and unlike the other laws, under the EAWR you are presumed guilty and have to establish your innocence by invoking the Defence Regulation 29. Perhaps some explanation is needed here. Each of the 16 Regulations has a status, in that it is either *absolute* or *reasonably practicable*.

Regulations that are *absolute* must be conformed to at all cost, whereas those that are *reasonably practicable* are conformed to provided that all reasonable steps have been taken to ensure safety. For the contravention of an *absolute* requirement, Regulation 29 is available as a defence in the event of criminal prosecution, provided the accused can demonstrate that they took all reasonable and diligent steps to prevent danger or injury.

No one wants to end up in court accused of negligence, and so we need to be sure that we know what we are doing when we are inspecting and testing.

THE BUILDING REGULATIONS PART 'P'

The following material is taken from *The Building Regulations 2000 approved document P*. © Crown Copyright material is reproduced with the permission of the controller of HMSO and Queen's Printer for Scotland.

The following are some details of Part 'P' of the Building Regulations (Table 1.1 and Figure 1.1).

Table 1.1 Examples of Work Notifiable and Not Notifiable

Notifiable (YES)	Not notifiable (NO)	Not applicable (N/A)
Examples of work	**Location A** **Within kitchens, bath/** **shower room, gardens,** **swimming/paddling** **pools & hot air saunas**	**Location B** **Outside of Location A**
A complete new installation or rewire	YES	YES
Consumer unit change	YES	YES
Installing a new final circuit (e.g. for lighting, socket outlets, a shower or a cooker)	YES	YES
Fitting and connecting an electric shower to an existing wiring point	YES	N/A
Adding a socket outlet to an existing final circuit	YES	NO
Adding a lighting point to an existing final circuit	YES	NO
Adding a fused connection unit to an existing final circuit	YES	NO
Installing and fitting a storage heater including final circuit	YES	YES
Installing extra low-voltage lighting (other than pre-assembled CE marked sets)	YES	YES
Installing a new supply to a garden shed or other building	YES	N/A
Installing a socket outlet or lighting point in a garden shed or other detached outbuilding	YES	N/A
Installing a garden pond pump, including supply	YES	N/A
Installing an electric hot air sauna	YES	N/A
Installing a solar photovoltaic power supply	YES	YES
Installing electric ceiling or floor heating	YES	YES

(Continued)

Table 1.1   Examples of Work Notifiable and Not Notifiable—Cont'd

Notifiable (YES)	Not notifiable (NO)	Not applicable (N/A)
Examples of work	**Location A** **Within kitchens, bath/** **shower room, gardens,** **swimming/paddling** **pools & hot air saunas**	**Location B** **Outside of Location A**
Installing an electricity generator	YES	YES
Installing telephone or extra low-voltage wiring and equipment for communications, information technology, signalling, control or similar purposes	YES	NO
Installing a socket outlet or lighting point outdoors	YES	YES
Installing or upgrading main or supplementary equipotential bonding	NO	NO
Connecting a cooker to an existing connection unit	NO	NO
Replacing a damaged cable for a single circuit, on a like-for-like basis	NO	NO
Replacing a damaged accessory, such as a socket outlet	NO	NO
Replacing a lighting fitting	NO	NO
Providing mechanical protection to an existing fixed installation	NO	NO
Fitting and final connection of storage heater to an existing adjacent wiring point	NO	NO
Connecting an item of equipment to an existing adjacent connection point	NO	NO
Replacing an immersion heater	NO	NO
Installing an additional socket outlet in a motor caravan	N/A	N/A

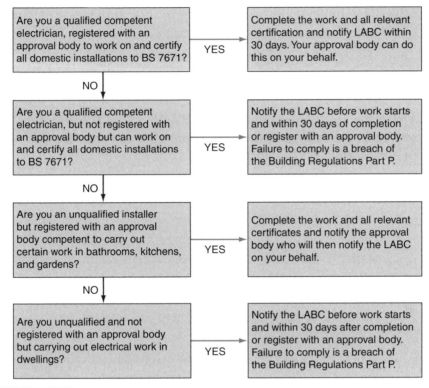

FIGURE 1.1 Compliance with Part 'P'.

Certification of notifiable work

a. Where the installer is registered with a Part P competent person self-certification scheme

1.18 Installers registered with a Part P competent person self-certification scheme are qualified to complete BS 7671 installation
certificates and should do so in respect of every job they under-
take. A copy of the certificate should always be given to the
person ordering the electrical installation work.

1.19 Where installers are registered with a Part P competent person self-certification scheme, a Building Regulations compliance certificate must be issued to the occupant either by the
installer or by the installer's registration body within 30 days of
the work being completed. The relevant building control body
should also receive a copy of the information on the certificate
within 30 days.

1.20 The Building Regulations call for the Building Regulations compliance certificate to be issued to the occupier. However, in the case of rented properties, the certificate may be sent to the person ordering the work with a copy sent also to the occupant.

b. Where the installer is not registered with a Part P competent person self-certification scheme but qualified to complete BS 7671 installation certificates

1.21 Where notifiable electrical installer work is carried out by a person not registered with a Part P competent person self-certification scheme, the work should be notified to a building control body (the local authority or an approved inspector) before work starts. Where the work is necessary because of an emergency the building control body should be notified as soon as possible. The building control body becomes responsible for making sure the work is safe and complies with all relevant requirements of the Building Regulations.

1.22 Where installers are qualified to carry out inspection and testing and completing the appropriate BS 7671 installation certificate, they should do so. A copy of the certificate should then be given to the building control body. The building control body will take this certificate into account in deciding what further action (if any) needs to be taken to make sure that the work is safe and complies fully with all relevant requirements. Building control bodies may ask for evidence that installers are qualified in this case.

1.23 Where the building control body decides that the work is safe and meets all building regulation requirements it will issue a building regulation completion certificate (the local authority) on request or a final certificate (an approved inspector).

c. Where installers are not qualified to complete BS 7671 completion certificates

1.24 Where such installers (who may be contractors or DIYers) carry out notifiable electrical work, the building control body must be notified before the work starts. Where the work is necessary because of an emergency the building control body should be notified as soon as possible. The building control body then becomes responsible for making sure that the work is safe and complies with all relevant requirements in the Building Regulations.

1.25 The amount of inspection and testing needed is for the building control body to decide based on the nature and extent of the electrical work. For relatively simple notifiable jobs, such as adding a socket outlet to a kitchen circuit, the inspection and testing requirements will be minimal. For a house re-wire, a full set of inspection and tests may need to be carried out.

1.26 The building control body may choose to carry out the inspection and testing itself, or to contract out some or all of the work to a special body which will then carry out the work on its behalf. Building control bodies will carry out the necessary inspection and testing at their expense, not at the householders' expense.

1.27 A building control body will not issue a BS 7671 installation certificate (as these can be issued only by those carrying out the work), but only a Building Regulations completion certificate (the local authority) or a final certificate (an approved inspector).

Third party certification

1.28 Unregistered installers should not themselves arrange for a third party to carry out final inspection and testing. The third party – not having supervised the work from the outset – would not be in a position to verify that the installation work complied fully with BS 7671:2008 requirements. An electrical installation certificate can be issued only by the installer responsible for the installation work.

1.29 A third party could only sign a BS 7671:2008 Periodic Inspection Report or similar. The Report would indicate that electrical safety tests had been carried out on the installation which met BS 7671:2008 criteria, but it could not verify that the installation complied fully with BS 7671:2008 requirements – for example with regard to routing of hidden cables.

Part 'P'

Part 'P' of the Building Regulations requires that certain electrical instal-lation work in domestic dwellings be certified and notified to the Local Authority Building Control (LABC). Failure to provide this notification may result in substantial fines.

Some approval bodies offer registration for all electrical work in domestic premises, these are known as full scope schemes (FS); other bodies offer registration for certain limited work in special locations such as kitchens, bathrooms, gardens, etc., these are known as defined scope schemes (DS).

In order to achieve and maintain competent person status, all approval bodies require an initial and thereafter annual registration fee and inspection visit.

Approval bodies (full scope FS and defined scope DS)		
NICEIC	(FS) & (DS)	0870 013 0900
NAPIT	(FS) & (DS)	0870 444 1392
ELESCA	(FS) & (DS)	0870 749 0080
BSI	(FS)	01442 230 442
BRE	(FS)	0870 609 6093
CORGI	(DS)	01256 392 200
OFTEC	(DS)	0845 658 5080

Apart from the knowledge required competently to carry out the verifica-tion process, the person conducting the inspection and test must be in possession of test instruments appropriate to the duty required of them.

INSTRUMENTS

In order to fulfil the basic requirements for testing to BS 7671, the follow-ing instruments are needed:

1. A low-resistance ohmmeter (continuity tester).
2. An insulation resistance tester.
3. A loop impedance tester.
4. A residual current device (RCD) tester.

5. A prospective fault current (PFC) tester.
6. An approved test lamp or voltage indicator.
7. A proving unit.
8. An earth electrode resistance tester.

Many instrument manufacturers have developed dual or multi-function instruments; hence it is quite common to have continuity and insulation resistance in one unit, loop impedance and PFC in one unit, loop impedance, PFC and RCD tests in one unit, etc. However, regardless of the various combinations, let us take a closer look at the individual test instrument requirements.

Low-resistance ohmmeters/continuity testers

Bells, buzzers, simple multimeters, etc., will all indicate whether or not a circuit is continuous, but will not show the difference between the resistance of, say, a 10 m length of 10 mm² conductor and a 10 m length of 1 mm² conductor. I use this example as an illustration, as it is based on a real experience of testing the continuity of a 10 mm² main protective bonding conductor between gas and water services. The services, some 10 m apart, were at either ends of a domestic premises. The 10 mm² conductor, connected to both services, disappeared under the floor, and a measurement between both ends indicated a resistance higher than expected. Further investigation revealed that just under the floor at each end, the 10 mm² conductor had been terminated in a connector block and the join between the two, about 8 m, had been wired with a 1 mm² conductor. Only a milli-ohmmeter would have detected such a fault.

A low-resistance ohmmeter should have a no-load source voltage of between 4 and 24 V, and be capable of delivering an ac or dc short circuit current of not less than 200 mA. It should have a resolution (i.e. a detectable difference in resistance) of at least 0.01 mV.

Insulation resistance testers

An *insulation resistance test* is the correct term for this form of testing, not a *megger* test as megger is a manufacturer's trade name, not the name of the test.

> An insulation resistance tester must be capable of delivering 1 mA when the required test voltage is applied across the minimum acceptable value of insulation resistance.

Hence, an instrument selected for use on a low-voltage (50 V ac–1000 V ac) system should be capable of delivering 1mA at 500V across a resistance of 1MΩ.

Loop impedance tester

This instrument functions by creating, in effect, an earth fault for a brief moment, and is connected to the circuit via a plug or by 'flying leads' connected separately to line, neutral and earth.

> The instrument should only allow an earth fault to exist for a maximum of 40ms, and a resolution of 0.01V is adequate for circuits up to 50A. Above this circuit rating, the ohmic values become too small to give such accuracy using a standard instrument, and more specialized equipment may be required.

RCD tester

Usually connected by the use of a plug, although 'flying leads' are needed for non-socket outlet circuits, this instrument allows a range of out-of-balance currents to flow through the RCD to cause its operation within specified time limits.

> The test instrument should not be operated for longer than 2s, and it should have a 10 per cent accuracy across the full range of test currents.

Earth electrode resistance tester

This is a 3 or 4 terminal, battery powered, resistance tester. Its application is discussed in Chapter 4.

PFC tester

This is either part of a combined PFC/Loop tester or a multi-function instrument. It is used to measure Prospective Short Circuit Current (PSCC) line to neutral or, Prospective Earth Fault Current (PEFC) line to earth.

Approved test lamp or voltage indicator

A flexible cord with a lamp attached is not an approved device, nor for that matter is the ubiquitous 'testascope' or 'neon screwdriver', which encourages the passage of current, at low voltage, through the body!

A typical approved test lamp is as shown in Figure 1.2.

The Health and Safety Executive, Guidance Note GS 38, recommends that the leads and probes associated with test lamps, voltage indicators, voltmeters, etc., have the following characteristics:

1. The leads should be adequately insulated and, ideally, fused.
2. The leads should be easily distinguished from each other by colour.
3. The leads should be flexible and sufficiently long for their purpose.
4. The probes should incorporate finger barriers, to prevent accidental contact with live parts.
5. The probes should be insulated and have a maximum of 2 mm of exposed metal, but preferably have spring loaded enclosed tips.

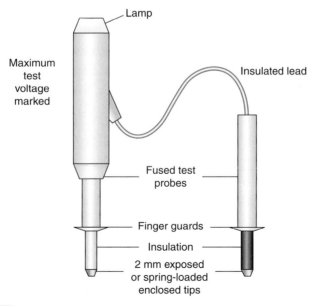

FIGURE 1.2 Approved test lamp.

Proving unit

This is an optional item of test equipment, in that test lamps should be proved on a known live supply which could, of course, be an adjacent socket or lighting point, etc. However, to prove a test lamp on such a known live supply may involve entry into enclosures with the associated hazards that such entry could bring. A proving unit is a compact device not much larger than a cigarette packet, which is capable of electronically developing 230 V dc across which the test lamp may be proved. The exception to this are test lamps incorporating 230 V lamps which will not activate from the small power source of the proving unit.

Test lamps must be proved against a voltage similar to that to be tested. Hence, proving test lamps that incorporate an internal check, that is shorting out the probes to make a buzzer sound is not acceptable if the voltage to be tested is higher than that delivered by the test lamp.

Care of test instruments

The EAWR (1989) requires that all electrical systems, this includes test instruments, be maintained to prevent danger. This does not restrict such maintenance to just a yearly calibration, but requires equipment to be kept in good condition in order that it is safe to use at all times. In consequence it is important to ensure the continual accuracy of instruments by comparing test readings against known values. This is most conveniently achieved by the use of 'checkboxes' which are readily available.

Whilst test instruments and associated leads, probes and clips, etc., used in the electrical contracting industry are robust in design and manufacture, they still need treating with care and protecting from mechanical damage. Keep test gear in a separate box or case away from tools and sharp objects and always check the general condition of a tester and leads before they are used.

Questions

1. State the statutory document most relevant to inspection and testing.
2. What is the minimum short circuit current to be delivered by a low resistance ohmmeter?
3. What current must be delivered by an insulation resistance tester when used at 500 V across a resistance of $1\,M\Omega$?
4. State the two tests carried out by a PFC tester.
5. What is the maximum length of exposed tip on the leads of a voltage indicator.

Answers

1. Electricity at Work Regulations (1989).
2. 200 mA.
3. 1 mA.
4. Prospective short circuit current (PSCC) and prospective earth fault current (PEFC).
5. 2 mm.

CHAPTER 2

Initial Verification

Important terms/topics covered by this chapter.

- Initial verification documentation
- Sequence of tests
- Inspection check list

By the end of this chapter the reader should,

- know the correct sequence of tests to be carried out,
- be aware of the information required by an inspector,
- be aware of the extent of the inspections required.

CIRCUMSTANCES WHICH REQUIRE AN INITIAL VERIFICATION

New installations or additions or alterations.

GENERAL REASONS FOR INITIAL VERIFICATION

1. To ensure equipment and accessories are to a relevant standard.
2. To prove compliance with BS 7671.
3. To ensure that the installation is not damaged so as to impair safety.

INFORMATION REQUIRED

Assessment of general characteristics sections 311, 312 and 313 together with information such as drawings, charts, etc., in accordance with Regulation 514.9.1 (see BS 7671:2008).

IEE Wiring Regulations. DOI: 10.1016/B978-0-08-096610-6.10002-8

DOCUMENTATION REQUIRED AND TO BE COMPLETED

Electrical Installation Certificate (EIC) signed or authenticated for the design and construction and then for the inspection and test (could be the same person). A schedule of test results and a schedule of inspections must accompany an EIC.

SEQUENCE OF TESTS

The IEE Regulations indicate a preferred sequence of tests and state that if, due to a defect, compliance cannot be achieved, the defect should be rectified and the test sequence started from the beginning. The tests for 'Protection by separation' and 'Insulation of non-conducting floors and walls' all require specialist equipment and in consequence will not be discussed here. The sequence of tests for an initial inspection and test is as follows:

1. Continuity of protective conductors.
2. Continuity of ring final circuit conductors.
3. Insulation resistance.
4. Protection against direct contact by barriers or enclosures.
5. Polarity.
6. Earth electrode resistance.
7. Earth fault loop impedance.
8. Additional protection (RCDs).
9. Prospective fault current between live conductors and to earth.
10. Phase sequence.
11. Functional testing.
12. Voltage drop. (Not normally required for initial verification)

BS 7671:2008 requires tests 1–5 be carried out in that order before the installation is energised and if there is an earth electrode, its testing should be included. It does not require the live tests 7–11 to follow a sequence and item 12 is not usually required for an initial verification.

Even though no sequence is specified, it would always be appropriate to conduct test 7 before test 8 as mal-functioning of an RCD under test could pose a shock risk if the loop impedance is inadequate.

One other test not included in Part 6 of the IEE Regulations but which nevertheless has to be carried out is external earth fault loop impedance (Z_e).

Before any testing is carried out, a detailed physical inspection must be made to ensure that all equipment is to a relevant British or Harmonized European Standard, that it is erected/installed in compliance with the IEE Regulations, and that it is not damaged such that it could cause danger. In order to comply with these requirements, the Regulations give a checklist of some 17 items that, where relevant, should be inspected.

However, before such an inspection, and test for that matter, is carried out, certain information *must* be available to the verifier. This information is the result of the assessment of fundamental principles BS 7671 Section 131 and the Assessment of General Characteristics required by IEE Regulations Part 3, sections 311, 312, 313, and drawings, charts and similar information relating to the installation. It is at this point that most readers who work in the real world of electrical installation will be lying on the floor laughing hysterically.

Let us assume that the designer and installer of the installation are competent professionals, and all of the required documentation is available.

Interestingly, one of the items on the checklist *is* the presence of diagrams, instructions and similar information. If these are missing then there is a deviation from the Regulations.

Another item on the list is the verification of conductors for current carrying capacity and voltage drop in accordance with the design. How on earth can this be verified without all the information? A 30A Type B circuit breaker (CB) or Type 2 miniature circuit breaker (MCB) protecting a length of 4 mm² conductor may look reasonable, but is it correct, and are you prepared to sign to say that it is unless you are sure? Let us look then at the general content of the checklist.

1. **Connection of conductors:** Are terminations electrically and mechanically sound? Is insulation and sheathing removed only to a minimum to allow satisfactory termination?
2. **Identification of conductors:** Are conductors correctly identified in accordance with the Regulations?

3. *Routing of cables:* Are cables installed such that account is taken of external influences such as mechanical damage, corrosion, heat, etc.?

4. *Conductor selection:* Are conductors selected for current carrying capacity and voltage drop in accordance with the design?

5. *Connection of single pole devices:* Are single pole protective and switching devices connected in the line conductor only?

6. *Accessories and equipment:* Are all accessories and items of equipment correctly connected?

7. *Thermal effects:* Are fire barriers present where required and protection against thermal effects provided?

8. *Protection against shock:* What methods have been used to attain both basic protection and fault protection?

9. *Mutual detrimental influence:* Are wiring systems installed such that they can have no harmful effect on non-electrical systems, or those systems of different currents or voltages are segregated where necessary?

10. *Isolation and switching:* Are there appropriate devices for isolation and switching correctly located and installed?

11. *Undervoltage:* Where undervoltage may give rise for concern, are there protective devices present?

12. *Labelling:* Are all protective devices, switches (where necessary) and terminals correctly labelled?

13. *External influences:* Have all items of equipment and protective measures been selected in accordance with the appropriate external influences?

14. *Access:* Are all means of access to switchgear and equipment adequate?

15. *Notices and signs:* Are danger notices and warning signs present?

16. *Diagrams:* Are diagrams, instructions and similar information relating to the installation available?

17. *Erection methods:* Have all wiring systems, accessories and equipment been selected and installed in accordance with the requirements of the Regulations, and are fixings for equipment adequate for the environment?

So, now that we have inspected all relevant items, and provided that there are no defects that may lead to a dangerous situation when testing, we can start the actual testing procedure.

Questions

1. An installation is to have the following tests conducted: (1) loop impedance, (2) polarity, (3) ring circuit continuity, and (4) insulation resistance. What is the correct sequence for carrying out the tests?
2. Which test is not normally required for an initial verification?
3. The details of which sections of BS 7671 are required to be made available to a person carrying out inspection and testing of an installation?
4. What inspection checklist item relates to damage to cables?

Answers

1. (3), (4), (2), (1).
2. Voltage drop.
3. 131, 311, 312, 313.
4. Routing of cables.

CHAPTER 3

Testing Continuity of Protective Conductors (Low-Resistance Ohmmeter)

Important terms/topics covered by this chapter:

- Protective bonding conductors
- Circuit protective conductors
- Parallel earth paths
- $(R_1 + R_2)$ values

By the end of this chapter the reader should,

- know what test instrument to use,
- understand the importance of disconnecting protective conductors for testing,
- know the importance of isolation, where protective conductors cannot be disconnected,
- be aware of the effects of parallel earth paths,
- be able to determine the approximate value of a protective conductor, given its length,
- know the preferred method of cpc continuity testing,
- know why $(R_1 + R_2)$ values are important.

All protective conductors, including main protective and supplementary bonding conductors, must be tested for continuity using a low-resistance ohmmeter.

For main protective bonding conductors there is no single fixed value of resistance above which the conductor would be deemed unsuitable. Each measured value, if indeed it is measurable for very short lengths, should be compared with the relevant value for a particular conductor length and size. Such values are shown in Table 3.1.

IEE Wiring Regulations. DOI: 10.1016/B978-0-08-096610-6.10003-X

Table 3.1   Resistance (in Ω) of Copper Conductors at 20°C

CSA (mm²)	Length (m)									
	5	10	15	20	25	30	35	40	45	50
1.0	0.09	0.18	0.27	0.36	0.45	0.54	0.63	0.72	0.82	0.9
1.5	0.06	0.12	0.18	0.24	0.3	0.36	0.43	0.48	0.55	0.6
2.5	0.04	0.07	0.11	0.15	0.19	0.22	0.26	0.3	0.33	0.37
4.0	0.023	0.05	0.07	0.09	0.12	0.14	0.16	0.18	0.21	0.23
6.0	0.02	0.03	0.05	0.06	0.08	0.09	0.11	0.13	0.14	0.16
10.0	0.01	0.02	0.03	0.04	0.05	0.06	0.063	0.07	0.08	0.09
16.0	0.006	0.01	0.02	0.023	0.03	0.034	0.04	0.05	0.05	0.06
25.0	0.004	0.007	0.01	0.015	0.02	0.022	0.026	0.03	0.033	0.04
35.0	0.003	0.005	0.008	0.01	0.013	0.016	0.019	0.02	0.024	0.03

Where a supplementary bonding conductor has been installed between *simultaneously accessible* exposed and extraneous conductive parts as an addition to fault protection and there is doubt as to the effectiveness of the equipotential bonding, then the resistance (R) of the conductor must be equal to or less than $50/I_a$. So, $R \leq 50/I_a$ where 50 is the voltage above which exposed metalwork should not rise, and I_a is the minimum current causing operation of the circuit protective device within 5 s.

For example, suppose a 45 A BS 3036 fuse protects a cooker circuit, the disconnection time for the circuit cannot be met, and so a supplementary bonding conductor has been installed between the cooker case and an adjacent central heating radiator. The resistance (R) of that conductor should not be greater than $50/I_a$, and I_a in this case is 145 A (see Figure 3.2B of the IEE Regulations), that is, $50/145 = 0.34\,\Omega$.

How then do we conduct a test to establish continuity of main or supplementary bonding conductors? Quite simple really, just connect the leads from a low-resistance ohmmeter to the ends of the bonding conductor (Figure 3.1). One end should be disconnected from its bonding clamp, otherwise any measurement may include the resistance of parallel paths of other earthed metalwork. Remember to zero/null the instrument first or, if this facility is not available, record the resistance of the test leads so that this value can be subtracted from the test reading.

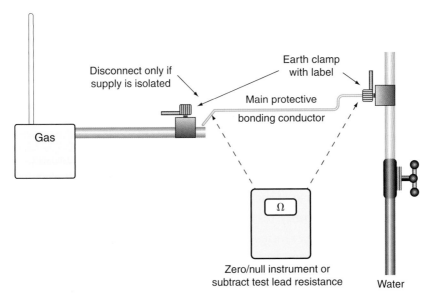

Disconnect only if
supply is isolated

Earth clamp
with label

Main protective
bonding conductor

Ω

Zero/null instrument or
subtract test lead resistance

Gas

Water

FIGURE 3.1   Testing main protective bonding.

Important Note

If the installation is in operation, then *never* disconnect protective bonding conductors unless the supply can be isolated. Without isolation, persons and livestock are at risk of electric shock. In this instance, or where the connections to extraneous conductive parts are not accessible, the test is conducted either between the connected bonding conductors or between extraneous conductive parts. The resistance value obtained should be no greater than $0.05\,\Omega$.

The continuity of circuit protective conductors may be established in the same way, but a second method is preferred, as the results of this second test indicate the value of $(R_1 + R_2)$ for the circuit in question.

The test is conducted in the following manner:

1. Temporarily link together the line conductor and cpc of the circuit concerned in the distribution board or consumer unit.
2. Test between line and cpc at each outlet in the circuit. A reading indicates continuity.
3. Record the test result obtained at the furthest point in the circuit. This value is $(R_1 + R_2)$ for the circuit, and is important for use with

the formula $Z_s = Z_e + (R_1 + R_2)$ for confirming measured values of Z_s or for calculation where Z_s cannot be measured. It should also be noted that for lighting circuits the test should be carried out at the switches as these are the furthest point for each luminaire.

Figure 3.2 illustrates the above method.

There may be some difficulty in determining the $(R_1 + R_2)$ values of circuits in installations that comprise steel conduit and trunking, and/or steel-wire armoured (SWA) and mineral-insulated metal-sheathed (MIMS) cables, because of the parallel earth paths that are likely to exist. In these cases, continuity tests may have to be carried out at the installation stage before accessories are connected or terminations made off as well as after completion.

Although it is no longer considered good working practice to use steel conduit or trunking as a protective conductor, it is permitted, and hence its continuity must be proved. The enclosure must be inspected along its length to ensure that it is sound and then the standard low-resistance test is performed.

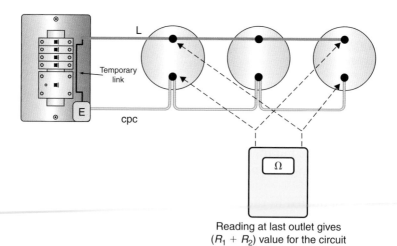

Reading at last outlet gives
$(R_1 + R_2)$ value for the circuit

FIGURE 3.2   Testing cpc continuity.

Questions

1. What instrument is used for testing the continuity of protective conductors?
2. What would be the approximate resistance value of a $10\,mm^2$ protective bonding conductor, $15\,m$ long?
3. What may be the effect on a resistance test reading taken between the connected ends of a protective bonding conductor?
4. Where, on a lighting circuit should a cpc continuity test be conducted?
5. What is the significance of the reading at the end of the circuit in Q.4 above?
6. Why is a value of $(R_1 + R_2)$ important, other than confirming cpc continuity?

Answers

1. Low resistance ohmmeter.
2. $0.03\ \Omega$.
3. A lower value of resistance than the actual conductor value due to parallel earth paths.
4. At all points on the circuit.
5. It is $(R_1 + R_2)$ for the circuit.
6. It can be used in the formula $Z_s = Z_e + (R_1 + R_2)$ to confirm a measured value of Z_s or to calculate a Z_s value where measurement is not possible.

CHAPTER 4

Testing Continuity of Ring Final Circuit Conductors (Low-Resistance Ohmmeter)

Important terms/topics covered by this chapter:

- Low resistance ohmmeter
- Ring final circuit interconnections
- Spurs
- $(R_1 + R_2)$ values
- Interpretation of test values

By the end of this chapter the reader should,

- know the reasons for conducting a ring final circuit continuity test,
- understand the problems that interconnections may create,
- understand why initial conductor cross-connections are made for the test,
- know how incorrect initial cross-connections are revealed,
- know why L to cpc values for flat sheathed cables vary slightly during the test,
- be able to interpret test results.

There are two main reasons for conducting this test:

1. To ensure that the ring circuit conductors are continuous, and indicate the value of $(R_1 + R_2)$ for the ring.
2. To establish that interconnections in the ring do not exist.

What then are interconnections in a ring circuit, and why is it important to locate them? Figure 4.1 shows a ring final circuit with an interconnection.

The most likely cause of the situation shown in Figure 4.1 is where a DIY enthusiast has added sockets P, Q, R and S to an existing rings A, B, C, D, E and F.

IEE Wiring Regulations. DOI: 10.1016/B978-0-08-096610-6.10004-1

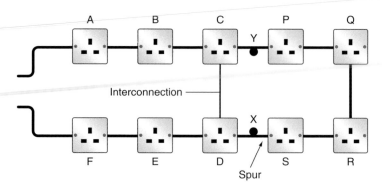

FIGURE 4.1   Ring circuit with an interconnection

In itself there is nothing wrong with this. The problem arises if a break occurs at, say, point Y, or the terminations fail in socket C or P. Then there would be four sockets all fed from the point X which would then become a spur.

So, how do we identify such a situation with or without breaks at point 'Y'? A simple resistance test between the ends of the line, neutral or circuit protective conductors will only indicate that a circuit exists, whether there are interconnections or not. The following test method is based on the philosophy that the resistance measured across any diameter of a perfect circle of conductor will always be the same value (Figure 4.2).

The perfect circle of conductor is achieved by cross-connecting the line and neutral legs of the ring (Figure 4.3). The test procedure is as follows:

1. Identify the opposite legs of the ring. This is quite easy with sheathed cables, but with singles, each conductor will have to be identified, probably by taking resistance measurements between each one and the closest socket outlet. This will give three high readings and three low readings thus establishing the opposite legs.
2. Take a resistance measurement between the ends of each conductor loop. Record this value.
3. Cross-connect the opposite ends of the line and neutral loops (Figure 4.4).

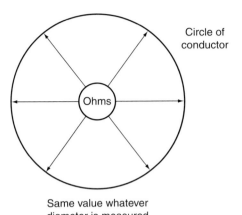

Circle of conductor

Ohms

Same value whatever diameter is measured

FIGURE 4.2 Measurement across diameter of a circle.

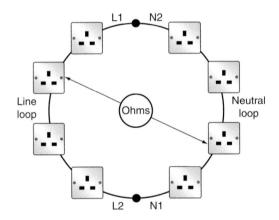

L1 N2

Line loop

Ohms

Neutral loop

L2 N1

FIGURE 4.3 Measurement across diameter of a ring circuit.

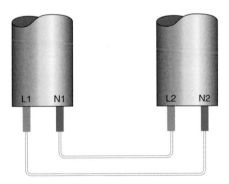

L1 N1 L2 N2

FIGURE 4.4 Ring circuit cross-connections L–N.

4. Measure between line and neutral at each socket on the ring. The readings obtained should be, for a perfect ring, substantially the same, and approximately ½ of the reading of individual loops. If an interconnection existed such as shown in Figure 4.1, then sockets A–F would all have similar readings, and those beyond the interconnection would have gradually increasing values to approximately the mid point of the ring, then decreasing values back towards the interconnection. If a break had occurred at point Y then the readings from socket S would increase to a maximum at socket P. One or two high readings are likely to indicate either loose connections or spurs. A null reading, that is an open circuit indication, is probably a reverse polarity, either line- or neutral-cpc reversal. These faults would clearly be rectified and the test at the suspect socket(s) is repeated. If the reading increases dramatically to the centre of the ring and then decreases again, it is likely that incorrect initial cross-connections of the legs of the ring have been made at Step 3.

5. Repeat the above procedure, but in this case cross-connect the line and cpc loops (Figure 4.5).

In this instance, if the cable is of the flat twin type, the readings at each socket will increase very slightly and then decrease around the ring. This difference, due to the line and cpc being different sizes, will not be significant enough to cause any concern. The measured value is very important; it is $R_1 + R_2$ for the ring.

As before, loose connections, spurs and, in this case, L–N cross polarity will be picked up.

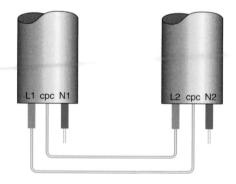

FIGURE 4.5   Ring circuit cross-connections L–cpc.

The details in Table 4.1 are typical approximate ohmic values for a healthy 70 m ring final circuit wired in 2.5 mm²/1.5 mm² flat twin and cpc cable. (In this case the cpc will be approximately 1.673 the L or N resistance.)

As already mentioned, null readings may indicate a reverse polarity. They could also indicate twisted conductors not in their terminal housing. The examples shown in Figure 4.6 may help to explain these situations.

Table 4.1 Resistance value for a 70 m ring circuit.

	L1–L2	N1–N2	cpc 1–cpc 2
Initial measurements	0.52	0.52	0.86
Reading at each socket	0.26	0.26	0.32–0.34
For spurs, each metre in length will add the following resistance to the above values	0.015	0.015	0.02

Socket	L–N Reading	L–cpc Reading
A	OK	No Reading
B	No Reading	OK
C	No Reading	No Reading

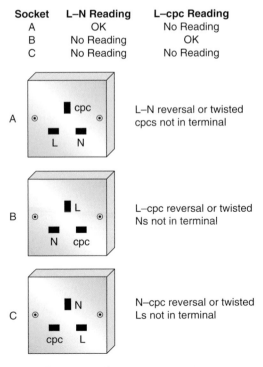

A — L–N reversal or twisted cpcs not in terminal

B — L–cpc reversal or twisted Ns not in terminal

C — N–cpc reversal or twisted Ls not in terminal

FIGURE 4.6 Reasons for null readings

Questions

1. State the reasons for conducting a ring final circuit continuity test.
2. What instrument is to be used for the test in Q.1 above?
3. Why are interconnections in ring circuits unacceptable?
4. Why are the ends of circuit conductors cross-connected for test purposes?
5. What are the effects on test results of correct and incorrect initial conductor cross-connections?
6. What may a null reading at a socket outlet indicate?
7. What does the L–cpc reading at each socket outlet on a ring signify?
8. Why will the L–cpc readings increase slightly and then decrease around a ring circuit wired in flat sheathed cable?
9. A ring final circuit is wired in $2.5\,mm^2$ singles (L, N & cpc) in conduit. If each loop has an end-to-end value of $0.4\,\Omega$, what would be the approximate expected value of $(R_1 + R_2)$?

Answers

1. Ensuring the ring is continuous and with no interconnections, and to establish a value for $(R_1 + R_2)$.
2. A low resistance ohmmeter.
3. A break in the ring beyond an interconnection may leave two or more socket outlets on a spur.
4. To create a perfect circle of conductor, the resistance across any diameter of which will give the same value.
5. Correct cross-connections give the same reading at each socket outlet, incorrect will result in greatly increased and decreased readings around the ring.
6. Twisted or touching conductors not in the socket outlet terminal or a reverse polarity.
7. $(R_1 + R_2)$ for the ring.
8. Because the cpc is smaller in size than the line conductor.
9. $0.2\,\Omega$.

CHAPTER 5

Testing Insulation Resistance (Insulation Resistance Tester)

Important terms/topics covered by this chapter:

- Insulation resistance tester
- Parallel resistances
- Disconnection of equipment
- Test procedure
- Test values
- SELV, PELV and FELV circuits
- Surge protective devices

By the end of this chapter the reader should,

- be aware of why the test is required,
- know the test instrument to be used,
- understand that insulation is a measure of resistances in parallel,
- know between which conductors the measurements should be made,
- know the test voltages and minimum values of insulation resistance,
- be aware of the need to test on circuits/equipment that have been isolated,
- be aware of the reasons for disconnecting various items of equipment,
- be able to calculate overall values of insulation resistance given individual circuit values.

This is probably the most used and yet abused test of them all. Affectionately known as 'meggering', an *insulation resistance test* is performed in order to ensure that the insulation of conductors, accessories and equipment is in a healthy condition, and will prevent dangerous leakage currents between conductors and between conductors and earth. It also indicates whether any short circuits exist.

IEE Wiring Regulations. DOI: 10.1016/B978-0-08-096610-6.10005-3

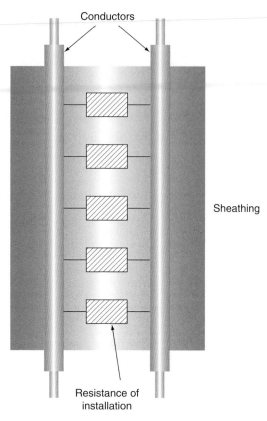

FIGURE 5.1 Parallel resistance of cable insulation.

Insulation resistance, as just discussed, is the resistance measured between conductors and is made up of countless millions of resistances in parallel (Figure 5.1).

The more resistances there are in parallel, the *lower* the overall resistance, and, in consequence, the longer a cable, the lower the insulation resistance. Add to this the fact that almost all installation circuits are also wired in parallel, and it becomes apparent that tests on large installations may, if measured as a whole, give pessimistically low values, even if there are no faults.

Under these circumstances, it is usual to break down such large installations into smaller sections, floor by floor, distribution circuit by distribution circuit, etc. This also helps, in the case of periodic testing, to minimize disruption. The test procedure is as follows:

1. Ensure the supply to the circuit/s in question is isolated.
2. Disconnect all items of equipment such as capacitors and indicator lamps as these are likely to give misleading results. Remove any items of equipment likely to be damaged by the test, such as dimmer switches, electronic timers, etc. Remove all lamps and accessories and disconnect fluorescent and discharge fittings. Ensure all fuses are in place, and circuit breakers and switches are in the on position. In some instances it may be impracticable to remove lamps, etc. and in this case the local switch controlling such equipment may be left in the off position. Where electronic devices cannot be disconnected, test only between lives and earth.
3. Join together all live conductors of the supply and test between this join and earth. Alternatively, test between each live conductor and earth in turn.
4. Test between line and neutral. For three phase systems, join together all lines and test between this join and neutral.
5. Then test between each of the lines. Alternatively, test between each of the live conductors in turn. Installations incorporating two-way lighting systems should be tested twice with the two-way switches in alternative positions.

Table 5.1 gives the test voltages and minimum values of insulation resistance for ELV and LV systems.

If a value of less than $2\,M\Omega$ is recorded it may indicate a situation where a fault is developing, but as yet still complies with the minimum permissible value. In this case each circuit should be tested separately in order to locate the problem.

In the case of SELV, PELV and electrical separation, Table 5.1 applies to their own circuit conductors. When they are with other circuits the

Table 5.1 Insulation resistance test requirements.

System	Test voltage	Minimum insulation resistance
SELV and PELV	250 V dc	0.5 MΩ
LV up to 500 V	500 V dc	1.0 MΩ
Over 500 V	1000 V dc	1.0 MΩ

insulation resistance between their conductors and those of the other circuits should be based on the highest voltage present. For FELV circuits the test voltage and the minimum value if insulation is the same as that for LV circuits up to 500 V i.e. 500 V dc and 1 MΩ.

Where surge protective devices exist, they should be disconnected. If this is not practicable the test voltage may be reduced to 250 V dc but the minimum value of insulation resistance remains at 1 MΩ.

Example 5.1

An installation comprising six circuits has individual insulation resistances of 2.5, 8, 200, 200, 200 and 200 MΩ, and so the total insulation resistance will be:

$$\frac{1}{R_t} = \frac{1}{2.5} + \frac{1}{8} + \frac{1}{200} + \frac{1}{200} + \frac{1}{200} + \frac{1}{200}$$
$$= 0.4 + 0.125 + 0.005 + 0.005 + 0.005 = 0.545$$

$$\frac{1}{R_t} = \frac{1}{0.545} = 1.83 \text{ M}\Omega$$

This is clearly greater than the 1.0 MΩ minimum but less than 2 MΩ. Had this value (1.83) been measured first, the circuits would need to have been investigated to identify the one/s that were suspect.

Note

It is important that a test for cpc continuity is conducted before an insulation resistance (IR) test. If a cpc was broken, and an IR test between line and cpc was carried out first, the result would be satisfactory, even if there was an L–cpc fault beyond the break. A subsequent cpc continuity test would reveal the break, which would be rectified, leaving an L–cpc fault undetected!!

Questions

1. What is the purpose of an insulation resistance test?
2. What instrument should be used?
3. Why do capacitors, neons, etc. need to be disconnected?
4. Why do items of electronic equipment need to be disconnected?
5. What action should be taken regarding switches and protective devices?
6. What is the test voltage and minimum value of insulation resistance for a 25 V FELV circuit?
7. What test voltage and minimum value of insulation resistance are appropriate for circuits incorporating surge protective devices?
8. Below what value of overall insulation resistance would an installation need to be investigated circuit by circuit?
9. Why may a large installation give a pessimistically low overall insulation resistance value?
10. What would be the total insulation resistance of an installation comprising circuits with the following values 3 MΩ, 12 MΩ, 100 MΩ and 150 MΩ?

Answers

1. To ensure that conductor insulation has not deteriorated or been damaged to an extent that excessive leakage currents can flow.
2. An insulation resistance tester.
3. To avoid misleading test results.
4. To avoid damage to such equipment.
5. All switches ON, all fuses IN, all circuit breakers ON.
6. 500 V dc; 1 MΩ.
7. 250 V dc; 1 MΩ.
8. 2 MΩ.
9. Because there are a large number of circuits all in parallel.
10. 2.3 MΩ.

CHAPTER 6

Special Tests

The next two tests are special in that they are not often required in the general type of installation. They also require special test equipment. In consequence, the requirements for these tests will only be briefly outlined in this short chapter.

PROTECTION BY BARRIERS OR ENCLOSURES

If, on site, basic protection is provided by fabricating an enclosure or erecting a barrier, it must be shown that the enclosure can provide a degree of protection of at least IPXXB or IP2X or, where required, at least IPXXD or IP4X.

An enclosure having a degree of protection IP2X can withstand the ingress of solid objects exceeding 12 mm diameter and fingers, IPXXB is protection against finger contact only. IP4X gives protection against solid objects and wires exceeding 1mm in diameter, IPXXD protects against wires exceeding 1mm in diameter only.

The test for IPXXB or IP2X is conducted with a 'standard test finger' which is supplied at a test voltage not less than 40 V dc and not more than 50 V dc. One end of the finger is connected in series with a lamp and live parts in the enclosure. When the end of the finger is introduced into the enclosure, provided the lamp does not light then the protection is satisfactory (Figure 6.1).

The test for IPXXD or IP4X is conducted with a rigid 1mm diameter wire with its end cut bent at right angles. Protection is afforded if the wire does not enter the enclosure.

PROTECTION BY NON-CONDUCTING LOCATION

This is a rare location and demands specialist equipment to measure the insulation resistance between insulated floors and walls at various points. Appendix 13 of BS 7671 outlines the tests required.

IEE Wiring Regulations. DOI: 10.1016/B978-0-08-096610-6.10006-5

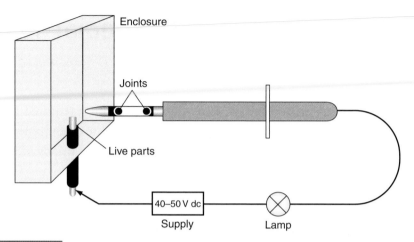

FIGURE 6.1 BS finger test.

CHAPTER 7

Testing Polarity (Low-Resistance Ohmmeter)

Important terms/topics covered by this chapter

- Edison screw lampholders
- Radial socket outlet circuits
- Supply polarity

By the end of this chapter the reader should,

- know the instrument to be used,
- know why BS EN 60238, E14 and E27 lampholders are exempt from polarity testing,
- know why ring final circuit polarity is not usually carried out during polarity testing,
- know how to check for line–cpc reversals on radial socket outlet circuits,
- know what live polarity test should be conducted.

This simple test, often overlooked, is just as important as all the others, and many serious injuries and electrocutions could have been prevented if only polarity checks had been carried out.

The requirements are:

1. All fuses and single pole switches and protective devices are in the line conductor.
2. The centre contact of an Edison screw type lampholder is connected to the line conductor (except E14 and 27 types to BS EN 60238, as these have threads of insulating material and the lamp must be fully inserted before L and N contacts are made).
3. All socket outlets and similar accessories are correctly wired.

IEE Wiring Regulations. DOI: 10.1016/B978-0-08-096610-6.10007-7

Although polarity is towards the end of the recommended test sequence, it would seem sensible, on lighting circuits, for example, to conduct this test at the same time as that for continuity of cpc's (Figure 7.1).

As discussed earlier, polarity on ring final circuit conductors is achieved simply by conducting the ring circuit test. For radial socket outlet circuits, however, this is a little more difficult. The continuity of the cpc will have already been proved by linking line and cpc and measuring between the same terminals at each socket. Whilst a line–cpc reversal would not have shown, a line–neutral reversal would, as there would have been no reading at the socket in question. This would have been remedied, and so only line–cpc reversals need to be checked. This can be done by linking together cpc and neutral at the origin and testing between the same terminals at each socket. A line–cpc reversal will result in no reading at the socket in question.

When the supply is connected, it is important to check that the incoming supply is correct. This is done using an approved voltage indicator at the intake position or close to it.

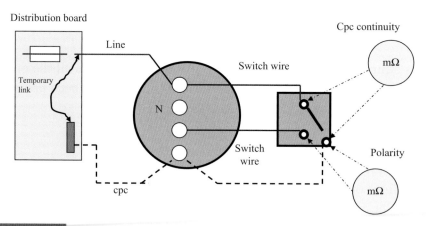

FIGURE 7.1 Lighting circuit polarity.

Questions

1. What instrument is used for testing polarity?
2. Why are BS EN 60238 E14 and E27 lampholders exempt from polarity testing?
3. At what point in a test sequence is the polarity of a ring final circuit checked?
4. How are line–cpc reversals identified in radial socket outlet circuits?
5. Where should live polarity tests be conducted?

Answers

1. Low resistance ohmmeter.
2. The lampholder screw thread is made of an insulating material.
3. When the ring final circuit continuity test is being conducted.
4. By cross-connecting neutral and cpc and testing between N and cpc at each socket.
5. At the supply intake to the installation.

CHAPTER 8

Testing Earth Electrode Resistance (Earth Electrode Resistance Tester or Loop Impedance Testers)

Important terms/topics covered by this chapter:

- Earth electrode resistance area
- Potential divider
- Current and potential electrodes
- Average value of earth electrode resistance
- Use of earth fault loop impedance tester

By the end of this chapter the reader should,

- know the test instruments that may be used,
- understand what is meant by the resistance area of an earth electrode,
- be able to state the electrodes involved when using an earth electrode resistance tester,
- know the extent of the resistance area of an electrode,
- know how to conduct a test using an earth electrode resistance tester,
- know what test may be conducted when the system is TT and is RCD protected,
- be able to determine the value of earth electrode resistance from test results.

In many rural areas, the supply system is TT and hence reliance is placed on the general mass of earth for a return path under earth fault conditions. Connection to earth is made by an electrode, usually of the rod type, and preferably installed as shown in Figure 8.1.

In order to determine the resistance of the earth return path, it is necessary to measure the resistance that the electrode has with earth. If we were to make such measurements at increasingly longer distances from the electrode, we would notice an increase in resistance up to about

IEE Wiring Regulations. DOI: 10.1016/B978-0-08-096610-6.10008-9

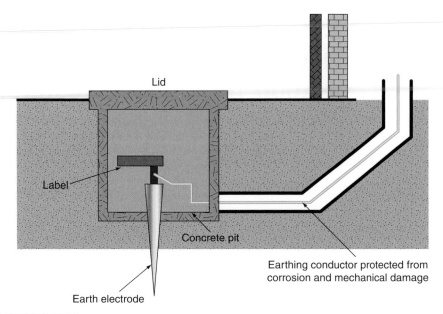

Lid

Label

Concrete pit

Earth electrode

Earthing conductor protected from
corrosion and mechanical damage

FIGURE 8.1   Earth electrode installation.

2.5–3 m from the rod, after which no further increase in resistance would
be noticed (Figure 8.2).

The maximum resistance recorded is the electrode resistance and the
area that extends to 2.5–3 m beyond the electrode is known as the earth
electrode resistance area.

There are two methods of making the measurement, one using a propri-
etary instrument and the other using a loop impedance tester.

METHOD 1: PROTECTION BY OVERCURRENT DEVICE

This method is based on the principle of the potential divider (Figure 8.3).

By varying the position of the slider the resistance at any point may be
calculated from $R = V/I$.

The earth electrode resistance test is conducted in a similar fashion with
the earth replacing the resistance and a potential electrode replacing the
slider (Figure 8.4). In Figure 8.4, the earthing conductor to the electrode
under test is temporarily disconnected.

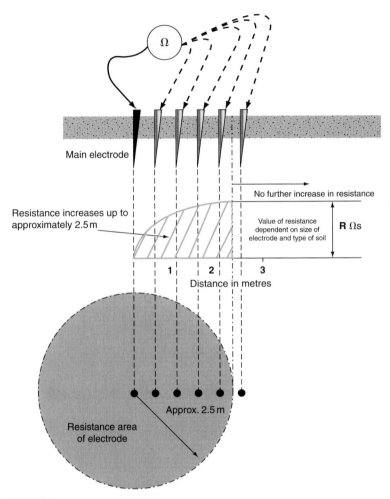

FIGURE 8.2   Earth electrode resistance area.

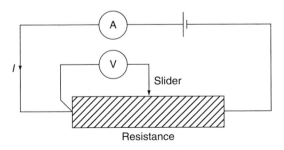

FIGURE 8.3   Potential divider.

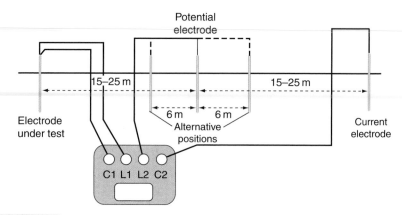

FIGURE 8.4   Earth electrode resistance test.

The method of test is as follows:

1. Place the current electrode (C2) away from the electrode under test, approximately 10 times its length (i.e. 30 m for a 3 m rod).
2. Place the potential electrode mid way.
3. Connect test instrument as shown.
4. Record resistance value.
5. Move the potential electrode approximately 6 m either side of the mid position, and record these two readings.
6. Take an average of these three readings (this is the earth electrode resistance).

For TT systems the result of this test will indicate compliance if the product of the electrode resistance and the operating current of the overcurrent device does not exceed 50 V. Clearly this will not be achieved when electrode resistances are high and hence will be more appropriate for electrodes used for earth connections for transformers and generators where the values need to be very small. Generally speaking the values obtained will result in the need for RCD protection.

METHOD 2: PROTECTION BY A RESIDUAL CURRENT DEVICE

In this case, an earth fault loop impedance test is carried out between the incoming line terminal and the electrode (a standard test for Z_e).

The value obtained is added to the cpc resistance of the protected circuits and this value is multiplied by the operating current of the RCD. The resulting value should not exceed 50 V. If it does, then Method 1 should be used to check the actual value of the electrode resistance.

Questions

1. What instruments may be used for earth electrode resistance testing?
2. What is the extent of the resistance area of an earth electrode?
3. For a 4 m electrode under test, at what distance away should the current electrode be placed?
4. Where should a potential electrode be initially placed when conducting an earth electrode resistance test?
5. Where are the alternative positions for the potential electrode?
6. What would be the resistance of an earth electrode if the test results gave values of 127 Ω, 129 Ω and 122 Ω?
7. What test may be performed when the system is TT and protected by an RCD?

Answers

1. Earth electrode resistance tester or earth fault loop impedance tester.
2. Approximately 2.5 m radius from the electrode.
3. 40 m minimum.
4. Centrally between the electrode under test and the current electrode.
5. 6 m either side of the potential electrode's initial position.
6. 126 Ω.
7. An earth fault loop impedance test.

CHAPTER 9

Testing Earth Fault Loop Impedance Tester

Important terms/topics covered by this chapter:

- Earth fault loop path
- Comparison of results with maximum values
- The rule of thumb
- RCD and cb operation
- Calculation of loop impedance
- External earth fault loop impedance

By the end of this chapter the reader should,

- know what instrument is required,
- be conversant with the various earth fault loop paths,
- know the test procedure,
- know how to adjust maximum values for comparison with test values,
- know to overcome the problems of RCD or cb operation during the test,
- be aware of the requirements for testing external earth fault loop impedance.

This is very important but, sadly, poorly understood. So let us remind ourselves of the component parts of the earth fault loop path (Figure 9.1). Starting at the point of fault:

1. The cpc.
2. The earthing conductor and main earthing terminal.
3. The return path via the earth for TT systems, and the metallic return path in the case of TN-S or TN-C-S systems. In the latter case the metallic return is the PEN conductor.
4. The earthed neutral of the supply transformer.
5. The transformer winding.
6. The line conductor back to the point of fault.

IEE Wiring Regulations. DOI: 10.1016/B978-0-08-096610-6.10009-0

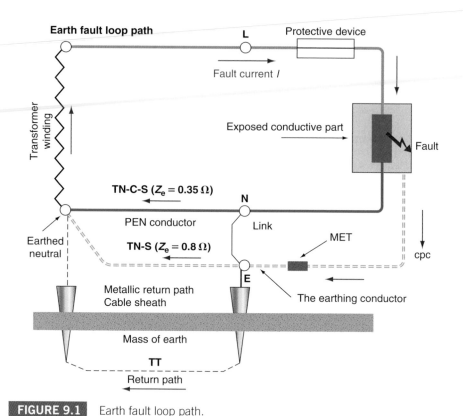

Earth fault loop path

L   Protective device

Fault current *I*

Exposed conductive part

Fault

Transformer winding

TN-C-S (*Z*ₑ = 0.35 Ω)

N

PEN conductor

Link

MET

Earthed neutral

TN-S (*Z*ₑ = 0.8 Ω)

E

cpc

Metallic return path
Cable sheath

The earthing conductor

Mass of earth

TT

Return path

FIGURE 9.1   Earth fault loop path.

Overcurrent protective devices must, under earth fault conditions, disconnect fast enough to reduce the risk of electric shock. This is achieved if the actual value of the earth fault loop impedance does not exceed the tabulated maximum values given in the relevant parts of the IEE Regulations.

The purpose of the test, therefore, is to determine the actual value of the loop impedance (Z_s), for comparison with those maximum values, and it is conducted as follows:

1. Ensure that all main equipotential bonding is in place.
2. Connect the test instrument either by its BS 1363 plug, or the 'flying leads', to the line, neutral and earth terminals at the remote end of the circuit under test. (If a neutral is not available, e.g. in the case of a three phase motor, connect the neutral probe to earth.)
3. Press to test and record the value indicated.

It must be understood that this instrument reading is *not valid for direct comparison with the tabulated maximum values*, as account must be taken of the ambient temperature at the time of test and the maximum conductor operating temperature, both of which will have an effect on conductor resistance. Hence, the $(R_1 + R_2)$ could be greater at the time of fault than at the time of test.

So, our measured value of Z_s must be corrected to allow for these possible increases in temperature occurring at a later date. This requires actually measuring the ambient temperature and applying factors in a formula.

Clearly this method of correcting Z_s is time consuming and unlikely to be commonly used. Hence, a rule of thumb method may be applied which simply requires that the measured value of Z_s does not exceed 0.8 of the appropriate tabulated maximum value. Table 9.1 gives the 0.8 values of tabulated loop impedance for direct comparison with measured values.

In effect, a loop impedance test places a line/earth fault on the installation, and if an RCD is present it may not be possible to conduct the test as it will keep tripping out. Unless the instrument can compensate for this, the value of Z_s will have to be calculated using the measured values of Z_e and $(R_1 + R_2)$ and the 0.8 rule applied. Remember, $Z_s = Z_e + (R_1 + R_2)$

EXTERNAL LOOP IMPEDANCE Z_E

The value of Z_e is measured at the origin of the installation on the supply side with the means of earthing disconnected, to avoid parallel paths. Do not conduct this test if the installation cannot be isolated.

Important Note

Never by-pass an RCD in order to conduct this test. Also, as this test creates a high current, some lower rated cb's may operate on overload. Do not replace with a higher rated breaker for test purposes, use the calculation method.

Table 9.1 Values of Loop Impedance for Comparison with Test Readings

Protection	Disconnection time		5A	6A	10A	15A	16A	20A
BS 3036 fuse	0.4 s	Z_s max	7.6	–	–	2.04	–	1.41
	5 s	Z_s max	14.16	–	–	4.2	–	3.06
BS 88 fuse	0.4 s	Z_s max	–	6.82	4.09	–	2.16	1.42
	5 s	Z_s max	–	10.8	5.94	–	3.33	2.32
BS 1361 fuse	0.4 s	Z_s max	8.36	–	–	2.62	–	1.36
	5 s	Z_s max	13.12	–	–	4	–	2.24
BS 1362 fuses	0.4 s	Z_s max	(3 A) 13.12			(13 A) 1.9	–	–
	5 s	Z_s max	(3 A) 18.56			(13 A) 3.06	–	–
BS 3871 MCB Type 1	0.4 & 5 s	Z_s max	9.2	7.6	4.6	3.06	2.87	2.3
BS 3871 MCB Type 2	0.4 & 5 s	Z_s max	5.25	4.37	2.62	1.75	1.64	1.31
BS 3871 MCB Type 3	0.4 & 5 s	Z_s max	3.68	3	1.84	1.22	1.15	0.92
BS EN 60898 CB Type B	0.4 & 5 s	Z_s max	(3 A) 12.26	6.13	3.68	–	2.3	1.84
BS EN 60898 CB Type C	0.4 & 5 s	Z_s max		3.06	1.84		1.15	0.92
BS EN 60898 CB Type D	0.4 & 5 s	Z_s max		1.54	0.92		0.57	0.46

Table 9.1 Values of Loop Impedance for Comparison with Test Readings—Con'd

25 A	30 A	32 A	40 A	45 A	50 A	60 A	63 A	80 A	100 A	125 A	160 A
–	0.87	–	–	–	–	–	–	–	–	–	–
–	2.11	–	–	1.27	–	0.89	–	–	0.42	–	–
1.15	–	0.83	–	–	–	–	–	–	–	–	–
1.84	–	1.47	1.08	–	0.83	–	0.67	0.45	0.33	0.26	0.2
–	0.92	–	–	–	–	–	–	–	–	–	–
–	1.47	–	–	0.79	–	0.56	–	0.4	0.29	–	–
–	–	–	–	–	–	–	–	–	–	–	–
–	–	–	–	–	–	–	–	–	–	–	–
1.84	1.53	1.44	1.15	1.02	0.92	–	0.73	–	–	–	–
1.05	0.87	0.82	0.67	0.58	0.52	–	0.42	–	–	–	–
0.74	0.61	0.57	0.46	0.41	0.37	–	0.29	–	–	–	–
1.47	–	1.15	0.92		0.74	–	0.58	0.46	0.37	0.3	–
0.75	–	0.57	0.46		0.37	–	0.288	023	0.18	0.15	–
0.37	–	0.288	0.23		0.18	–	0.14	0.12	0.09	0.07	–

Questions

1. What instrument is used for earth fault loop impedance testing?
2. Which earthing system includes a PEN conductor?
3. Before testing, what action should be taken regarding equipotential bonding?
4. Why is the 0.8 rule applied?
5. Is a measured value of loop impedance of $1.2\,\Omega$ satisfactory if the tabulated maximum value is $1.44\,\Omega$?
6. How may a value for loop impedance Z_s be obtained if an RCD or a cb operates when the test is conducted?
7. What action is required regarding the earthing conductor of an installation before conducting a test for external loop impedance Z_e?
8. Why is the action in Q7 above required and what other measure must be taken?

Answers

1. An earth fault loop impedance tester.
2. TN-C-S.
3. Ensure it is connected.
4. To compensate for increased ambient and conductor operating temperature.
5. No, as the corrected maximum would be $0.8 \times 1.44 = 1.15\,\Omega$.
6. Calculation from $Z_s = Z_e + (R_1 + R_2)$.
7. It must be disconnected.
8. To avoid parallel paths. The supply to the installation must be isolated.

CHAPTER 10

Additional Protection (RCD Tester)

Important terms/topics covered by this chapter:

- RCD/RCBO test requirements
- Uses for RCDs/RCBOs
- Determination of RCD/RCBO rating

By the end of this chapter the reader should,

- know what instrument should be used,
- know the test requirements for various type of RCD/RCBO,
- know the instrument settings required,
- be able to identify where RCDs/RCBOs are required,
- know how to determine the rating of RCDs/RCBOs.

RCD/RCBO OPERATION

Where RCDs and RCBOs are used as additional protection against shock, it is essential that they operate within set parameters. The RCD testers used are designed to do just this, and the basic tests required are as follows (Table 10.1):

1. Set the test instrument to the rating of the RCD.
2. Set the test instrument to half rated trip $(1/2I_{\Delta n})$.
3. Operate the instrument and the RCD should not trip.
4. Set the instrument to deliver the full rated tripping current of the RCD $(I_{\Delta n})$.
5. Operate the instrument and the RCD should trip out in the required time.
6. A 30 mA RCD or less, operating at $5 \times I_{\Delta n}$, should trip in 40 ms.

Note

Most RCD testers have the facility to test, separately, each half cycle of the supply and so each test should be done at 0° and 180°. The highest reading should be recorded.

IEE Wiring Regulations. DOI: 10.1016/B978-0-08-096610-6.10010-7

Table 10.1 RCD/RCBO Test requirements.		
RCD type	**Half rated**	**Full trip current**
BS 4239 and BS 7288 sockets	No trip	<200 ms
BS 4239 with time delay	No trip	½ time delay + 200 ms to time delay + 200 ms
BS EN 61009 or BS EN 61009 RCBO	No trip	<300 ms
As above but Type S with time delay	No trip	130 ms ≤ *I* ≤ 500 ms

Note

This last test is not required for RCDs rated over 30 mA.

There seems to be a popular misconception regarding the ratings and uses of RCDs in that they are the panacea for all electrical ills and the only useful rating is 30 mA!

Firstly, RCDs are not fail safe devices, they are electromechanical in operation and can malfunction. Secondly, general purpose RCDs are manufactured in ratings from 5 to 500 mA and have many uses. The accepted lethal level of shock current is 50 mA and hence RCDs rated at 30 mA or less would be appropriate for use where shock is an increased risk. The following list indicates the residual current ratings and uses of RCDs as stated in BS 7671.

REQUIREMENTS FOR RCD PROTECTION

30 mA

- All socket outlets rated at not more than 20 A and for unsupervised general use.
- Mobile equipment rated at not more than 32 A for use outdoors.
- All circuits in a bath/shower room.
- Preferred for all circuits in a TT system.
- All cables installed less than 50 mm from the surface of a wall or partition (in the safe zones) if the installation is unsupervised, and

also at any depth if the construction of the wall or partition includes metallic parts.

- In zones 0, 1 and 2 of swimming pool locations.
- All circuits in a location containing saunas, etc.
- Socket outlet final circuits not exceeding 32 A in agricultural locations.
- Circuits supplying Class II equipment in restrictive conductive locations.
- Each socket outlet in caravan parks and marinas and final circuit for houseboats.
- All socket outlet circuits rated not more than 32 A for show stands, etc.
- All socket outlet circuits rated not more than 32 A for construction sites (where reduced low voltage, etc. is not used).
- All socket outlets supplying equipment outside mobile or transportable units.
- All circuits in caravans.
- All circuits in circuses, etc.
- A circuit supplying Class II heating equipment for floor and ceiling heating systems.

100 mA

- Socket outlet final circuits of rating exceeding 32 A in agricultural locations.

300 mA

- At the origin of a temporary supply to circuses, etc.
- Where there is a risk of fire due to storage of combustible materials.
- All circuits (except socket outlets) in agricultural locations.

500 mA

- Any circuit supplying one or more socket outlets of rating exceeding 32 A, on a construction site.

Where loop impedance values cannot be met, RCDs of an appropriate rating can be installed. Their rating can be determined from

$$I_{\Delta n} = \frac{50}{Z_s}$$

where $I_{\Delta n}$ is the rated operating current of the device, 50 is the touch voltage, and Z_s is the measured loop impedance.

Questions

1. What test instrument is required for RCD/RCBO testing?
2. What is the maximum operating time for a BS EN 61008 RCD at full rated current?
3. What is the maximum operating time for a 30 mA RCD when tested at 150 mA?
4. What maximum rating of RCD should be used for a 63 A socket outlet on a construction site?
5. What rating of RCD is required for a caravan installation?
6. What would be the required maximum rating of an RCD where the earth fault loop impedance was 167 Ω?

Answers

1. An RCD tester.
2. 300 ms.
3. 40 ms.
4. 500 mA.
5. 30 mA.
6. 300 mA.

CHAPTER 11

Prospective Fault Current (PFC/PSCC Tester)

There is a requirement to determine the prospective fault current at the origin of an installation and at relevant points throughout. At the origin this may be ascertained by enquiry or measurement, whereas at other points, measurement is the only option.

Where the lowest rated protective device in the installation has a breaking capacity higher than the PFC at the origin, then measurement at other points is not needed.

PFC is a generic term and can be either prospective short circuit current, PSCC (between lines or line and neutral) or prospective earth fault current, PEFC (between line and earth). Both should be measured and the highest value recorded, although there is no harm in recording both.

The testers are designed for single phase use, so where a value of PSCC is required for a three phase system it may be determined by multiplying the single phase by 2, or more accurately 1.732.

IEE Wiring Regulations. DOI: 10.1016/B978-0-08-096610-6.10011-9

CHAPTER 12

Check of Phase Sequence (Phase Sequence Indicator)

For three phase systems it is important to have knowledge of the phase rotation of the supply and at various points within an installation. It is convention for this rotation to be normally Brown, Black, Grey or LI, L2, L3. The direction of three phase motors can be reversed simply by reversing any two phases. In consequence the correct sequence is essential to ensure the right rotation.

Paralleling of two three phase generators or of a generator to the three phase public supply system requires their phase sequences to be synchronized.

The instrument is simply a small three phase motor with a dial that indicates in which direction the supply is rotating (Figure 12.1).

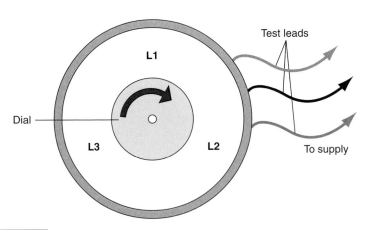

FIGURE 12.1 Phase sequence indicator.

IEE Wiring Regulations. DOI: 10.1016/B978-0-08-096610-6.10012-0

CHAPTER 13

Functional Testing

All RCDs have a built-in test facility in the form of a test button. Operating this test facility creates an artificial out-of-balance condition that causes the device to trip. This only checks the mechanics of the tripping operation, it does not check the condition of the electrical installation and hence is not a substitute for the tests discussed in Chapter 10.

There should be a notice in a prominent position at or near the origin of the installation where the device is located indicating that the test button should be operated quarterly. For temporary installations it is recommended that this operation be carried out at more regular intervals.

All other items of equipment such as switchgear, controlgear interlocks, etc., must be checked to ensure that they are correctly mounted and adjusted, and that they function correctly.

This could involve the operation of, for example, two-way switching, dimmer switches, timers, thermostats, main isolators, circuit breakers, etc.

IEE Wiring Regulations. DOI: 10.1016/B978-0-08-096610-6.10013-2

CHAPTER 14

Voltage Drop (Approved Voltmeter)

There may be a requirement to determine the voltage at the terminals of equipment to ensure that items of electrical equipment will function correctly and safely. BS 7671 is a little vague on exactly how this may be achieved, mentioning measuring circuit impedances or using calculations. The voltage drop is probably best established by measuring the voltage at the origin and at the end of a particular circuit, subtracting the two and ensuring the drop does not exceed the values indicated in Table 14.1 (it is not usually required to establish this during an initial verification); 3% for LV lighting circuits, 5% for LV power circuits, hence:

Table 14.1   Maximum permissible voltage drop.

	LV Lighting 3%	LV Power 5%
Single phase 230 V	6.9 V	11.5 V
Three phase 400 V	12 V	20 V

IEE Wiring Regulations. DOI: 10.1016/B978-0-08-096610-6.10014-4

CHAPTER 15

Periodic Inspection

Important terms/topics covered by this chapter:

- Circumstances requiring periodic inspection and testing
- General reasons for periodic inspection and testing
- Conditions to be investigated
- Documentation to be completed
- General areas of non-compliance that may be revealed

By the end of this chapter the reader should,

- know why periodic inspection and testing is required,
- know what general conditions within an installation need investigation,
- know what documentation needs to be completed,
- know what action is required if there is insufficient information/ drawings, etc.,
- understand the extent to which dismantling and sampling should take place,
- be aware of the conditions that may permit an installation to be exempt from periodic inspection and testing.

PERIODIC INSPECTION AND TESTING

This is the province of the experienced inspector who has not only the knowledge and technical expertise to competently carry out the testing process but who is also fully conversant with correct electrical installation practices.

CIRCUMSTANCES WHICH REQUIRE A PERIODIC INSPECTION AND TEST

Test and inspection is due; insurance, mortgage, licensing reasons; change of use; change of ownership; after additions or alterations; after damage; change of loading; to assess compliance with current regulations.

IEE Wiring Regulations. DOI: 10.1016/B978-0-08-096610-6.10015-6

GENERAL REASONS FOR A PERIODIC INSPECTION AND TEST

1. To ensure the safety of persons and livestock.
2. To ensure protection of property from fire and heat.
3. To ensure that the installation is not damaged so as to impair safety.
4. To ensure that the installation is not defective and complies with the current regulations.

GENERAL AREAS OF INVESTIGATION

Safety, Ageing, Damage, Corrosion, Overloading, Wear and tear, External influences and Suitability (as an *aide memoire* the author calls this his SADCOWES list!!).

DOCUMENTATION TO BE COMPLETED

Periodic inspection/condition report, a schedule of test results and a schedule of inspections.

SEQUENCE OF TESTS

There is no required sequence. However, the sequence for an Initial Verification is preferred if possible.

Periodic inspection and testing could be such a simple and straightforward process. Nevertheless it usually tends to be complicated and frustrating.

On the domestic scene, I doubt if any house owner actually decides to have a regular inspection, the comment being, 'If it works it must be OK'. It is usually only when there is a change of ownership that the mortgage companies insist on an electrical survey. The worst cases are, however, in industry and commerce.

Periodic inspections are often requested by clients, reluctantly, to satisfy insurers or an impending visit by the HSE. Even then it is usually the case that 'you can't turn that off' or 'why can't you just test this bit and then issue a certificate for the whole lot'.

Under the rare circumstances that an inspection and test is genuinely requested due to responsible concerns for the safety of staff, etc., it is difficult to convince the client that, as there are no drawings, or information about the installation, and that no switchgear is labelled, etc., you are going to be on site for a considerable time and at a considerable cost.

When there are no drawings or items of information, especially on a large installation, there may be a degree of exploratory work to be carried out in order to ensure safety whilst inspecting and testing. If it is felt that it may be unsafe to continue with the inspection and test, then drawings and information must be produced in order to avoid contravening Section 6 of the Health and Safety at Work Act.

However, let us assume (in our wildest dreams) that, as with an Initial Verification, the original installation was erected in accordance with the Wiring Regulations, and any alterations and/or additions have been faithfully recorded and all the original documentation/diagrams/charts, etc., are readily available!

A periodic inspection and test under these circumstances should be relatively easy, as little dismantling of the installation will be necessary, and the bulk of the work will be inspection.

Inspection should be carried out with the supply disconnected as it may be necessary to gain access to wiring in enclosures, etc., and hence with large installations it will probably need considerable liaison with the client to arrange convenient times for interruption of supplies to various parts of the installation.

This is also the case when testing protective conductors, as these must *never* be disconnected unless the supply can be isolated. It is particularly important in the case of main protective bonding conductors which need to be disconnected in order to measure Z_e.

In general an inspection should reveal:

1. Any aspects of the installation that may impair the safety of persons and livestock against the effects of electric shock and burns.
2. That there are no installation defects that could give rise to heat and fire and hence damage property.
3. That the installation is not damaged or deteriorated so as to impair safety.
4. That any defects or non-compliance with the Regulations, which may give rise to danger, are identified.

As was mentioned earlier, dismantling should be kept to a minimum as this process may create faults. Hence a certain amount of sampling will be required.

The amount of sampling would need to be commensurate with the number of defects being found.

From the testing point of view, not all of the tests carried out on the initial inspection may need to be applied. This decision depends on the condition of the installation.

The continuity of protective conductors is clearly important as are the tests for insulation resistance, loop impedance and operation of RCDs, but one wonders if polarity tests are necessary if the installation has remained undisturbed since the last inspection. The same applies to ring circuit continuity as the L–N test is applied to detect interconnections in the ring, which could not occur on their own.

It should be noted that if an installation is effectively supervised in normal use, then periodic inspection and testing can be replaced by regular maintenance by skilled persons. This would only apply to, say, factory installations where there are permanent maintenance staff.

Questions

1. State two circumstances that would result in the need for a periodic inspection?
2. State three installation conditions that may need investigation?
3. State the three items of documentation that will need to be completed?
4. When may exploratory work be required before commencing a periodic inspection and test of a large installation?
5. Why should dismantling be kept to a minimum?
6. When may periodic inspection and testing be replaced by routine maintenance?

Answers

1. Any two from, due date, mortgage, insurance, etc.
2. Any three from SADCOWES list.
3. Periodic/condition report.
4. Where there is a lack of information/drawings, etc., and it may be unsafe to continue without them.
5. To avoid causing damage and creating faults.
6. When the installation is under effective supervision and the maintenance is carried out by skilled persons.

CHAPTER 16

Certification

Having completed all the inspection checks and carried out all the relevant tests, it remains to document all this information. This is done on Electrical Installation Certificates, Electrical Installation Condition Reports (often referred to as Periodic Inspection Reports), schedules, test results, Minor Electrical Installation Works Certificates and any other documentation you wish to append to the foregoing. Examples of such documentation are shown in BS 7671 and the IEE Guidance Note 3 on inspection and testing.

This documentation is vitally important. It has to be correct and signed or authenticated by a competent person. Electrical Installation Certificates and Electrical Installation Condition Reports must be accompanied by a schedule of test results and a schedule of inspections for them to be valid. It should be noted that both Electrical Installation Certificates and Minor Electrical Installation Works Certificates should be signed or otherwise authenticated by competent persons in respect of the design, the construction and the inspection and testing of the installation. The Electrical Installation Condition Report is signed by the inspector.

(For larger installations there may be more than one designer, hence the certificate has space for two signatures, i.e. designer 1 and designer 2.) It could be, of course, that for a very small company, one person signs all three parts. Whatever the case, the original must be given to the person ordering the work, and a duplicate retained by the contractor.

One important aspect of an EIC is the recommended interval between inspections. This should be evaluated by the designer and will depend on the type of installation and its usage. In some cases the time interval is mandatory, especially where environments are subject to use by the public. The IEE Guidance Note 3 give recommended maximum frequencies between inspections.

An Electrical Installation Condition Report is very similar in part to an Electrical Installation Certificate in respect of details of the installation, i.e. maximum demand, type of earthing system, Z_e, etc. The rest of the form deals with the extent and limitations of the inspection and test, recommendations, and a summary of the installation. The record of the extent and limitations of the inspection is very important. It must be agreed with the client or other third party exactly what parts of the installation will be covered by the report and those that will not. The interval until the next test is determined by the inspector.

With regard to the schedule of test results, test values should be recorded unadjusted, any compensation for temperature, etc., being made after the testing is completed.

Any alterations or additions to an installation will be subject to the issue of an Electrical Installation Certificate, except where the addition is, say, a single point added to an existing circuit, then the work is subject to the issue of an MEIWC.

Summary:

1. The addition of points to existing circuits requires a Minor Electrical Installation Works Certificate.
2. A new installation or an addition or alteration that comprises new circuits requires an Electrical Installation Certificate.
3. An existing installation requires an Electrical Installation Condition Report.

Note

Points (2) and (3) must be accompanied by a schedule of test results and a schedule of inspections.

As the client/customer is to receive the originals of any certification, it is important that *all* relevant details are completed correctly. This ensures that future inspectors are aware of the installation details and test results which may indicate a slow progressive deterioration in some or all of the installation.

These certificates, etc., could also form part of a 'sellers pack' when a client wishes to sell a property.

The following is a general guide to completing the necessary documentation and should be read in conjunction with the examples given in BS 7671 and the IEE Guidance Note 3.

ELECTRICAL INSTALLATION CERTIFICATE

1. Details of client:
   Name: *Full name*
   Address: *Full address and post code*
   Description: *Domestic, industrial, commercial*
   Extent: *What work has been carried out (e.g. full re-wire, new shower circuit, etc.). Tick a relevant box.*

2. Designer/constructor/tester:
   Details of each or could be one person.
   Note: Departures are not faults, they are systems/equipment, etc., that are not detailed in BS 7671 but may be perfectly satisfactory.

3. Next test:
   When the next test should be carried out and decided by the designer.

4. Supply characteristics and earthing arrangements:
   Earthing system: *Tick relevant box (TT, TN-S, etc.)*
   Live conductors: *Tick relevant boxes*
   Nominal voltage: *Obtain from supplier, but usually 230V single phase U and U_0 but 400V U and 230 U_0 for three phase*
   Frequency: *From supplier but usually 50 Hz*
   PFC: *From supplier or measured. Supplier usually gives 16 kA*
   Z_e: *From supplier or measurement. Supplier usually gives 0.8V for TN-S; 0.35V for TN-C-S and 21V for TT systems*
   Main fuse: *Usually BS 1361, rating depends on maximum demand.*

5. Particulars of installation:
   Means of earthing: *Tick 'supplier's facility' for TN systems 'earth electrode' for TT systems*
   Maximum demand: *Value without diversity*
   Earth electrode: *Measured value or N/A.*
   (a) Earthing and bonding:
   Conductors: *Actual sizes and material, usually copper*

Main switch or circuit breaker (could be separate units or part of a consumer control unit): *BS number; Rating, current and voltage; Location; 'not address', that is where is it located in the building; Fuse rating if in a switch-fuse, else N/A; RCD details only if used as a main switch.*

6. Comments on existing installation:
   Write down any defects found in other parts of the installation which may have been revealed during an addition or an alteration.

7. Schedules:
   Indicate the number of test and inspection 'schedules that will accompany this certificate'.

PERIODIC/CONDITION REPORT

1. Details of client:
   Name: *Full name (could be a landlord, etc.)*
   Address: *Full address and post code (may be different to the installation address)*
   Purpose: *For example, due date; change of owner/tenant; change of use, etc.*

2. Details of installation:
   Occupier: *Could be the client or a tenant*
   Installation: *Could be the whole or part (give details)*
   Address: *Full and post code*
   Description: *Tick relevant box*
   Age: *If not known, say so, or educated guess*
   Alterations: *Tick relevant box and insert age where known*
   Last inspection: *Insert date or 'not known'*
   Records: *Tick relevant box.*

3. Extent and limitations:
   Full details of what is being tested (extent) and what is not (limitations). If not enough space on form add extra sheets.

4. Next inspection:
   Filled in by inspector and signed, etc., under declaration.

5. Supply details:
   As per an Electrical Installation Certificate.

6. Observations:
Tick relevant box, if work is required, record details and enter relevant code (1, 2, 3 or 4) in space on right-hand side.

7. Summary:
Comment on overall condition. Only common sense and experience can determine whether satisfactory or unsatisfactory.

8. Schedules:
Attach completed schedules of inspections and test results.

MINOR ELECTRICAL INSTALLATION WORKS CERTIFICATE

Only to be used when simple additions or alterations are made, *not when a new circuit is added.*

1. Description: Full description of work
Address: *Full address*
Date: *Date when work was carried out*
Departures: *These are not faults, they are systems/equipment, etc. that are not detailed in BS 7671 but may be perfectly satisfactory (this is usually N/A).*

2. Installation details:
Earthing: *Tick a relevant box*
Method of fault protection: *99% of the time this will be automatic disconnection of supply*
Protective device: *Enter type and rating. For example, BS EN 60898 CB type B, 20A*
Comments: *Note any defects/faults/omissions in other parts of the installation seen while conducting the minor works.*

3. Tests:
Earth continuity: *Measured and then tick in box if OK*
Insulation resistance: *Standard tests and results*
EFLI (Z_s): *Standard test and results*
Polarity: *Standard tests then tick in box if OK*
RCD: *Standard tests, record operating current and time.*

4. Declaration
Name, address, signature, etc.

CONTENTS OF A TYPICAL SCHEDULE OF TEST RESULTS

1. Contractor:	*Full name of tester.*
2. Date:	*Date of test.*
3. Signature:	*Signature of tester.*
4. Method of fault protection:	*99% Automatic disconnection of supply but could be SELV, etc.*
5. Vulnerable equipment:	*Dimmers, electronic timers, CH controllers, etc. (i.e. anything electronic).*
6. Address:	*Full, or if in a large installation, the location of a particular DB.*
7. Earthing:	*Indicate TT, TN-S, TN-C-S etc.*
8. Z_e at origin:	*Measured value.*
9. PFC:	*Record the highest value that is PEFC or PSCC (should be the same for TN-C-S).*
10. Confirmation of supply polarity:	*Tick box.*
11. Instruments:	*Record serial numbers of each instrument, or one number for a composite instrument.*
12. Description:	*Suggest initial or periodic or whatever part of the installation is involved. For example, initial verification on a new shower circuit.*
13. kVA rating of protection:	*Taken from device.*
14. Type and rating:	*For example, BS EN 60898 CB type B, 32A, or BS 88 40A, etc.*
15. Wiring conductors:	*Size of live and cpc, e.g. 2.5 mm^2 /1.5 mm^2.*
16. Test results:	*Enter all measured results, not corrected values.*

SCHEDULE OF INSPECTIONS (AS PER BS 7671)

Do not leave boxes uncompleted: N/A in a box if it is not relevant; ✓ in a box if it has been inspected and is OK; X in a box if it has been inspected and is incorrect.

Characteristic Curves and Maximum Loop Impedance Values for BS 3871 Miniature Circuit Breakers

Maximum earth fault loop impedances (Z_s) for BS 3871 miniature circuit breakers for 0.4 and 5 s with U_o of 230 V.

TYPE 1

Rating (A)	5	6	10	15	16	20	25	30	32	40	45	50	63	I_n
Z_s (Ω)	11.5	9.58	5.75	3.83	3.6	2.87	2.3	1.9	1.8	1.44	1.27	1.15	0.91	$57.5/I_n$

TYPE 2

Rating (A)	5	6	10	15	16	20	25	30	32	40	45	50	63	I_n
Z_s (Ω)	6.6	5.47	3.29	2.19	2.05	1.64	1.31	1.09	1.03	0.82	0.73	0.66	0.52	$33/I_n$

TYPE 3

Rating (A)	5	6	10	15	16	20	25	30	32	40	45	50	63	I_n
Z_s (Ω)	4.6	3.83	2.3	1.53	1.44	1.15	0.92	0.77	0.72	0.57	0.51	0.45	0.36	$23/I_n$

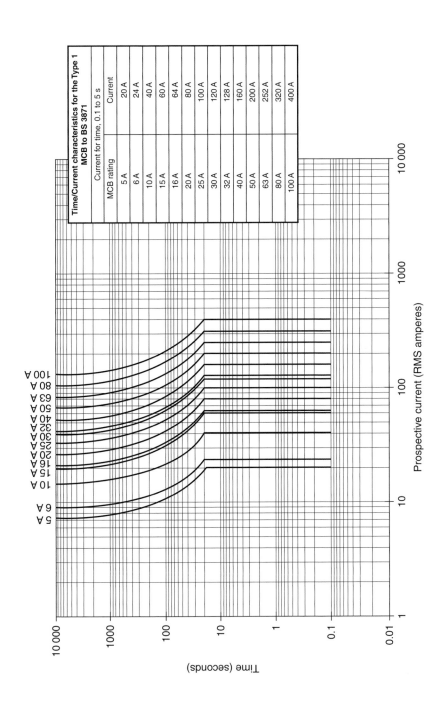

Time/Current characteristics for the Type 1 MCB to BS 3871

MCB rating	Current for time, 0.1 to 5 s
	Current
5 A	20 A
6 A	24 A
10 A	40 A
15 A	60 A
16 A	64 A
20 A	80 A
25 A	100 A
30 A	120 A
32 A	128 A
40 A	160 A
50 A	200 A
63 A	252 A
80 A	320 A
100 A	400 A

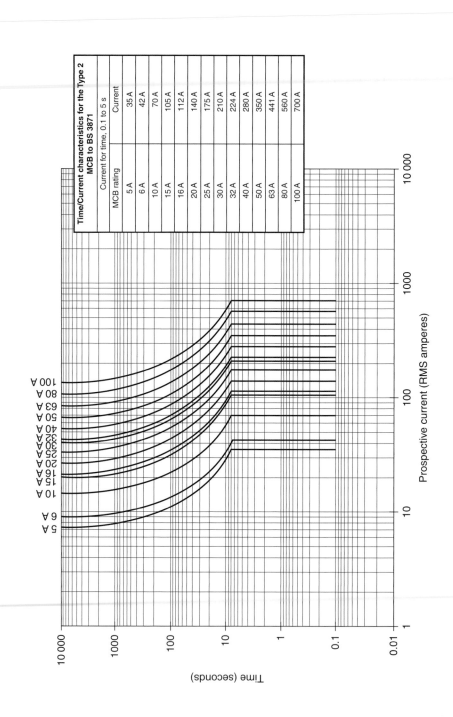

Time/Current characteristics for the Type 2 MCB to BS 3871

MCB rating	Current for time, 0.1 to 5 s
	Current
5 A	35 A
6 A	42 A
10 A	70 A
15 A	105 A
16 A	112 A
20 A	140 A
25 A	175 A
30 A	210 A
32 A	224 A
40 A	280 A
50 A	350 A
63 A	441 A
80 A	560 A
100 A	700 A

Prospective current (RMS amperes)

Time (seconds)

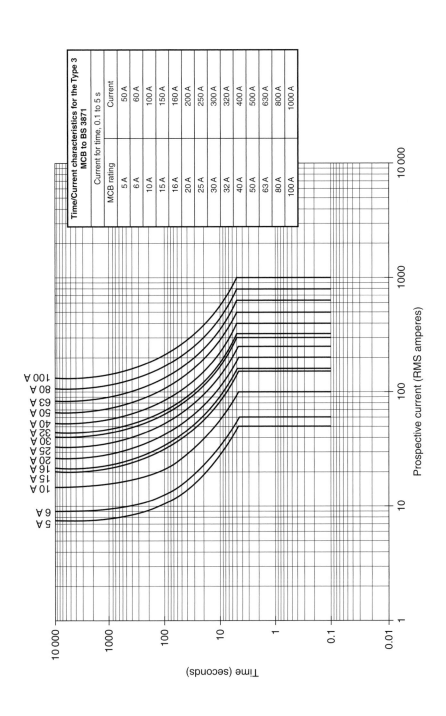

Time/Current characteristics for the Type 3 MCB to BS 3871

MCB rating	Current for time, 0.1 to 5 s Current
5 A	50 A
6 A	60 A
10 A	100 A
15 A	150 A
16 A	160 A
20 A	200 A
25 A	250 A
30 A	300 A
32 A	320 A
40 A	400 A
50 A	500 A
63 A	630 A
80 A	800 A
100 A	1000 A

Prospective current (RMS amperes)

Time (seconds)

Sample Paper

SECTION A – SHORT ANSWER

1. State the three reasons for carrying out an initial verification of a new installation.
2. There are various documents that are relevant to the inspection and testing of an installation, state:
   (a) one statutory item of documentation
   (b) two non-statutory items of documentation.
3. An Electrical Installation Certificate should be accompanied by signed documentation regarding three stages of an installation. What are these stages?
4. Apart from wear and tear, state three areas of investigation that you would consider when carrying out a periodic inspection and test of an installation.
5. State three human senses that could be used during an inspection of an installation.
6. During a test on an installation, the following readings were obtained: $20\,M\Omega$; $8\,kA$; $22\,ms$. List the instruments which gave these readings.
7. The following circuits are to be tested for insulation resistance. State the test voltages to be applied and the minimum acceptable value of insulation resistance in each case:
   (a) SELV circuit
   (b) LV circuit up to $500\,V$
   (c) LV circuit over $500\,V$.
8. List the first three tests that should be carried out during an initial verification on a new domestic installation.
9. The test for the continuity of a cpc in a radial circuit feeding one socket outlet uses a temporary link and a milli-ohmmeter, state:
   (a) where the temporary link is connected
   (b) where the milli-ohmmeter is connected
   (c) what the meter reading represents.
10. List three different protective conductors that would need to be connected to the main earthing terminal of an installation.

11. The following readings were obtained during the initial tests on a healthy ring final circuit:

L1 – L2 – 0.8 Ω; N1 – N2 – 0.8 Ω; cpc1 – cpc2 – 0.8 Ω

   (a) what readings would you expect:
      (i) between L and N conductors at each socket outlet?
      (ii) between L and cpc at each socket outlet?
      (iii) what the L to cpc reading represents?

12. What happens to:
   (a) conductor resistance when conductor length increases?
   (b) insulation resistance when cable length increases?
   (c) conductor resistance when conductor area increases?

13. List three precautions to be taken prior to commencing an insulation resistance test on an installation.

14. An enclosure with an accessible horizontal top surface is to be used as a means of basic protection. State the IP codes with which the enclosure should at least comply.

15. What degree of protection is offered by enclosures offering the following:
   (a) IPXXB
   (b) IP4X
   (c) IPX8.

16. List three reasons for conducting a dead polarity test on an installation.

17. What earthing systems are attributed to the following:
   (a) supply with the only earth located at the supply transformer?
   (b) a multicore supply cable with a separate neutral and earth?
   (c) a supply cable in which the functions of earth and neutral are performed by one conductor?

18. State three locations where special considerations should be made with regards to electrical installations.

19. From the formula

$$Z_s = Z_e + \frac{(R_1 + R_2) \times 12 \times L}{1000}$$

   (a) what is represented by:
      (i) Z_e?

(ii) R_2?

(iii) 12?

20. State any three functional tests that may be carried out on a domestic installation.

SECTION B

Figure A2.1 shows the layout of the electrical installation in a new detached garage. You are to carry out an initial verification of that installation.

21. (a) What documentation/information will you require in order to carry out the verification?

(b) Where should it be located?

(c) What particularly important details regarding this installation should have been included on such documentation?

(d) What consideration should be given to the existing installation from which this new installation is fed?

22. List five areas of inspection for this installation that should be carried out prior to testing.

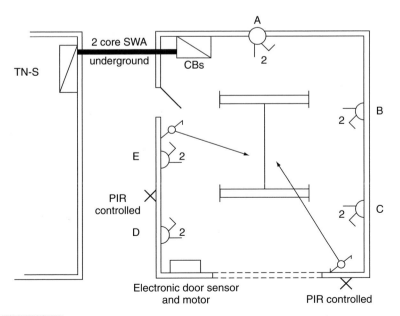

FIGURE A2.1 Installation layout.

23. The following test results were obtained from a ring final circuit continuity test. State if the readings for each socket are satisfactory and give reasons for those readings you feel are unsatisfactory.

Line, neutral and cpc loops $= 0.5\ \Omega$

Socket	L–N	L–cpc
A	0.25	0.26
B	No reading	0.25
C	0.35	0.24
D	0.24	No reading
E	0.26	0.26

24. (a) Describe in detail how you would carry out an insulation resistance test on this installation.
   (b) The test result indicates an overall value of $1.75\ M\Omega$, what actions, if any, should be taken. Explain your reasons.

25. A loop impedance test on the lighting circuit cannot be conducted, as the 6 A Type B circuit breaker keeps tripping out. Explain why this is, and how the problem may be overcome in order to conduct the test.

26. (a) The electronic door sensor/motor is wired on its own radial circuit, list all the component parts of the earth fault loop path associated with this circuit in the event of a fault to earth.
   (b) If the maximum value of loop impedance for this circuit is $2.4\ \Omega$ and an earth fault causes a current of 115 A, show by calculation if this value will disconnect the circuit in the required time.

Suggested Solutions to Sample Paper

SECTION A

1. - to ensure accessories, etc., to relevant standard
   - to ensure compliance with BS 7671
   - to ensure no damage that may cause danger.
   (3 marks)

2. (a) *The electricity at work regulations* (1 mark)
   (b) any two of:
   - BS 7671
   - *Guidance Note 3*
   - *The On-site Guide*.
   (2 marks)

3. - design
   - construction
   - inspection and testing.
   (3 marks)

4. Safety, ageing, damage, corrosion, overload, external influences and suitability
   - to ensure safety of persons and livestock
   - to ensure protection from fire and heat
   - to ensure that the installation is not damaged so as to impair safety
   - to ensure that the installation is not defective and complies with current regulations.
   (3 marks)

5. - visual
   - touch
   - smell.
   (3 marks)

6. - insulation resistance tester
   - prospective short circuit current tester
   - RCD tester.
   (3 marks)

7. ▪ 250 V and 0.5 MΩ
   ▪ 500 V and 1.0 MΩ
   ▪ 1000 V and 1.0 MΩ.
   (3 marks)

8. ▪ continuity of protective conductors
   ▪ continuity of ring final circuit conductors
   ▪ insulation resistance.
   (3 marks)

9. ▪ between L and E at the consumer unit
   ▪ between L and E at the socket outlet
   ▪ this value is $(R_1 + R_2)$ for the circuit.
   (3 marks)

10. Any three of:
    ▪ circuit protective conductor
    ▪ main protective bonding conductor
    ▪ earthing conductor
    ▪ lightning conductor.
    (3 marks)

11. ▪ 0.4 Ω
    ▪ 0.4 Ω
    ▪ $(R_1 + R_2)$ for the ring.
    (3 marks)

12. ▪ increases
    ▪ decreases
    ▪ decreases.
    (3 marks)

13. Any three of:
    ▪ check on existence of electronic equipment
    ▪ check there are no neons, capacitors, etc., in circuit
    ▪ all switches closed and accessories equipment removed
    ▪ no danger to persons or livestock by conducting the test.
    (3 marks)

14. Any three from:
    ▪ IP2X
    ▪ IPXXB
    ▪ IP4X

- IPXXD.
(3 marks)

15. • finger contact only
  - small foreign solid bodies or 1 mm diameter wires
  - total submersion.
  (3 marks)

16. • all single pole devices in line conductor only
  - centre contact of Edison screw lampholders in line conductor
  - all accessories correctly connected.
  (3 marks)

17. • TT
  - TN-S
  - TN-C-S.
  (3 marks)

18. Any three from Part 7 BS 7671:2008.
  (3 marks)

19. • external loop impedance
  - resistance of cpc
  - multiplier for conductor operating temperature.
  (3 marks)

20. Any three of:
  - test button operation of an RCD
  - operation of dimmer switch
  - operation of main isolating switch
  - operation of MCBs
  - operation of two-way switching.
  (3 marks)

SECTION B

21. (a) The results of the assessment of general characteristics sections 311, 312 and 313, and diagrams, charts and similar information regarding the installation.
  (5 marks)

  (b) In or adjacent to the distribution board.
  (3 marks)

(c) Reference to the electronic door sensor and the PIR controlled external luminaires as these could be vulnerable to a typical test.

(3 marks)

(d) Maximum demand, rating of consumer unit, earthing and bonding arrangements, capacity of main protective device, etc.

(4 marks)

22. Any relevant five from the BS 7671 Inspection schedule.

(3 marks each)

23. • Socket A OK as readings are approximately 1/2 of 0.5.

(3 marks)

   ▪ Socket B cross-polarity L–cpc, or twisted N conductors not in N terminal.

(3 marks)

   ▪ Socket C loose neutral connection.

(3 marks)

   ▪ Socket D cross-polarity L–N, or twisted cpcs not in terminal.

(3 marks)

   ▪ Socket E OK.

(3 marks)

24. (a) Conduct the test from the house as this will then include the SWA cable. Disconnect the supply to the PIR controlled lights and the electronic door sensor. Disconnect the capacitor and ballast at each fluorescent luminaire. With the garage main switch and the circuit breakers ON and any accessories unplugged, test at 500 V between live conductors connected together and earth and then between each live conductor. Operate the two-way switches during each test. The test readings should not be less than 1.0 MΩ.

(8 marks)

   (b) If any reading is below 2 MΩ, then there may be a latent defect and each circuit should be tested separately to locate any faults.

(7 marks)

25. As a loop impedance tester delivers a high current for a short time, it is not unusual for sensitive circuit breakers with low ratings to trip out on overload. The loop impedance in such cases will have to

be determined by a combination of measurement and calculation as follows:

Measure Z_e and measure $(R_1 + R_2)$ for the circuit, then $Z_s = Z_e + (R_1 + R_2)$.

(7 marks)

26. (a) ▪ The point of fault
    ▪ The cpc
    ▪ The steel wire armour of the garage supply
    ▪ The earthing conductor
    ▪ The metallic earth return path of the supply cable
    ▪ The earthed neutral of the transformer
    ▪ The transformer winding
    ▪ Line conductors.

(8 marks)

(b) $Z_s = \dfrac{U_0}{I_a} = \dfrac{230}{115} = 2\Omega$

so circuit protection will operate fast enough.

(7 marks)

Index